Energy use policies and carbon

Arun Advani

Institute for Fiscal Studies and University College London

Samuela Bassi

Grantham Research Institute on Climate Change and the Environment, and Centre for Climate Change Economics and Policy, LSE

Alex Bowen

Grantham Research Institute on Climate Change and the Environment, and Centre for Climate Change Economics and Policy, LSE

Sam Fankhauser

Grantham Research Institute on Climate Change and the Environment, and Centre for Climate Change Economics and Policy, LSE

Paul Johnson

Institute for Fiscal Studies

Andrew Leicester

Frontier Economics (work carried out whilst at the Institute for Fiscal Studies)

George Stoye

Institute for Fiscal Studies and University College London

Copy-edited by Emma Dain, T&T Productions Ltd

Acknowledgements

We are grateful to the Esmée Fairbairn Foundation (reference 11-2886) and the ESRC Centre for the Microeconomic Analysis of Public Policy (reference RES-544-28-5001) at the Institute for Fiscal Studies (IFS) for funding this work. The Grantham Research Institute on Climate Change and the Environment is supported by Grantham Foundation for the Protection of the Environment, and the ESRC through the Centre for Climate Change Economics and Policy. We would also like to thank Simeon Thornton, who acted as an advisor to this project, Ute Collier, Antoine Dechezleprêre, David Grover, Jenny Hill, Jim Skea, Michael Spackman and Dimitri Zenghelis for helpful comments on earlier drafts, and representatives from the Department for Energy and Climate Change and HM Treasury for useful discussions.

Contact details

arun_a@ifs.org.uk; s.bassi@lse.ac.uk; a.bowen1@lse.ac.uk; s.fankhauser@lse.ac.uk; paul_j@ifs.org.uk; andrew.leicester@frontier-economics.com; george_s@ifs.org.uk

Institute for Fiscal Studies (IFS), 7 Ridgmount Street, London WC1E 7AE; +44(0)20 7291 4800

Centre for Climate Change Economics and Policy and the Grantham Research Institute on Climate Change and the Environment, London School of Economics and Political Science (LSE), Houghton Street, London WC2A 2AE; +44(0)20 7107 5433

Disclaimer

Published by
The Institute for Fiscal Studies
7 Ridgmount Street
London WC1E 7AE
Tel: +44 (0) 20-7291 4800
Fax: +44 (0) 20-7323 4780
Email: mailbox@ifs.org.uk
Website: http://www.ifs.org.uk

in association with
London Publishing Partnership
http://www.londonpublishingpartnership.co.uk

978-1-909463-22-6

Contents

Executive Summary

Energy policy in the UK has many objectives – to ensure cheap and reliable supply, to avoid fuel poverty, to reduce carbon emissions and other pollutants and to raise revenue. Some of these objectives have been formalised in explicit statutory targets on fuel poverty, renewable energy and greenhouse gas (GHG) emissions.

Over the years, public policymakers have put in place an array of measures to achieve these objectives – we identify over 20 policy interventions aimed at, or felt directly by, energy users. There are many more that fall outside the scope of this report, including regulatory arrangements, which determine how competition in this sector works, and other policies which concern energy producers only. This has made UK energy use policy very complex and in places inefficient and inconsistent.

The aim of this report is to analyse and assess the policy landscape faced by UK energy users – both households and firms – and explore the potential for improvement. We highlight a number of reform options and recommendations for firm and household energy use policies. They include some changes that could be implemented quickly and more radical reforms that would require further consultation and time.

Economic principles for energy use policy

There are good economic reasons for government intervention in the energy market associated with environmental costs (or 'externalities') of energy, equity issues related to fuel poverty and perhaps market failures related to technological innovation and energy efficiency.

A key concern is the need to reduce GHG emissions. The government is committed to reducing economy-wide emissions by 80% by 2050, relative to 1990 levels, and energy use is central to this objective. We take this objective as given. The purpose of this analysis is not to question the objective, but rather to ask whether the objective is being pursued in an efficient manner.

Given an objective to reduce emissions, a central part of policy should be a single, consistent carbon price. The environmental damage associated with GHGs depends on the amount that is emitted, regardless of who creates these emissions or where they occur. Variation in the carbon price faced by different users is inefficient – though for a given target the appropriate price should rise over time.

Given that the UK government has set itself clear targets, the natural way to identify the 'right' carbon price is to set it at a level consistent with meeting those targets. This is the approach adopted by UK government analysts when valuing carbon emissions in policy appraisals, and it generates a value (in sectors of the economy not covered by the EU Emissions Trading Scheme) of £59/tCO_2e in 2013, rising to £65/tCO_2e in 2020. In fact, the required price for carbon in the electricity market may be higher than this, given the government's stated objective of largely decarbonising electricity by 2030.

While the underlying principle is clear, there are multiple objectives in energy policy, which may provide reasons for deviating from entirely uniform carbon prices:

- Energy use is a necessity, so policies that increase energy costs will have disproportionate effects on the welfare of poorer households. This may justify some combination of differential pricing, support for energy efficiency and direct income support for affected households.
- There may be a case for differential pricing to support particular technologies given market failures in innovation and uncertainty over optimal technology mixes in the future.
- If other countries are not pursuing similar policies, a consistent carbon price affecting energy-intensive companies operating in traded sectors may disadvantage those companies. This could result in carbon 'leakage' as carbon-intensive activity moves abroad, creating costs for the UK economy without reducing total emissions.

The presence of multiple objectives may therefore justify implicit carbon prices that are uneven. However, it is not obvious that all the government's objectives are expressed sensibly. It is doubtful, for example, whether we need separate objectives for fuel poverty as well as overall poverty, and for renewable energy as well as carbon emissions. And the government's explicit objective to increase the proportion of tax revenue accounted for by environmental taxes has been defined so restrictively – for example, excluding receipts from fuel duties – and in such a way that meeting it is trivial.

Variation in carbon prices

In fact, the layering of policies over time has led to a substantial variation in carbon prices across users and fuel types (illustrated in Figure ES.1). Gas attracts a much lower carbon price than electricity and residential consumers pay lower carbon prices than business.

The implicit carbon price for electricity use is much higher than for gas. This reflects a whole host of policies both 'upstream' (at the point of generation), such as the EU emissions trading system, and 'downstream' (at the point of use). The downstream policies differ between companies and households.

There are no policies imposing a carbon price on gas use by households.

The picture for households is heavily influenced by how one thinks about VAT. VAT on household energy consumption is paid at the reduced rate of 5% (compared with the standard 20% rate). That is an effective and substantial subsidy. Ignoring that, the effective carbon tax on household electricity consumption would be £59/tCO_2e – not far off that on most business and in line with the government's stated carbon price of £59/tCO_2e. Take account of that subsidy, however, and the effective carbon price on household electricity consumption falls to just £6/tCO_2e. For gas the figure is negative. This can be seen in Table ES.1, which shows implicit carbon prices faced by households in 2013 and 2020 before and after the impact of the implicit VAT subsidy is considered.

Figure ES.1: Implicit carbon prices (£/tCO_2e) by end-user and fuel type, 2013 and 2020

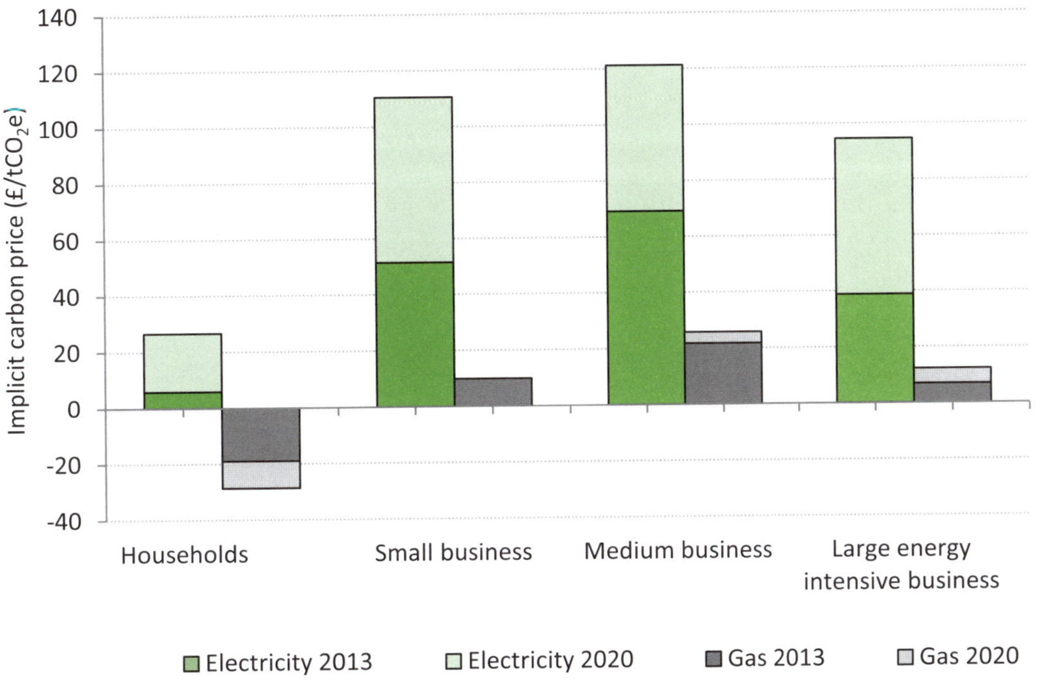

Source: Authors' calculations (see Appendix B for full sources and methodology). Figures in 2013 prices.

Table ES.1: Implicit carbon prices (£/tCO_2e) faced by households excluding and including the implicit VAT subsidy, by fuel type in 2013 and 2020

	Year	Implicit carbon price (excluding VAT subsidy)	Implicit VAT subsidy	Implicit carbon price
Electricity	2013	59	-53	6
	2020	126	-100	27
Gas	2013	13	-32	-19
	2020	13	-41	-29

Source and notes: Authors' calculations (see Appendix B for full sources and methodology). Figures may not sum precisely due to rounding. 2020 figures are given in 2013 prices.

To give a sense of scale, the VAT 'subsidy' reduces household energy bills by 14.3% (or £99 for the 'average' £691 gas bill and £82 for the 'average' electricity bill of £576), while the costs of various policies increase gas bills by 5% and electricity bills by 16% (or £33 for the average gas bill of £658 and £80 for the average electricity bill of £496 before policy costs are accounted for).

Governments have found it easier to impose taxes on firms than on households and as a result firms face higher carbon prices than households. But big, energy-intensive firms face a lower price than smaller, less energy-intensive ones. This reflects concerns about carbon 'leaking' across borders.

Going forward, the divergence in carbon prices between electricity and gas will only grow. The new electricity market arrangements will raise prices to support renewable generation. There are no plans to impose any carbon price on gas. The result is that (taking account of the VAT subsidy) households will face a carbon price of –£29/tCO_2e for gas and +£27/tCO_2e for electricity (2013–4 prices). This will not encourage switching from gas to (increasingly low carbon) electricity. Businesses will continue to face considerably higher prices than households.

Even accounting for the multiple social, environmental and economic objectives in the energy sector, this does not look like an optimal set of policies and moving even further away from a consistent carbon price suggests the situation is only going to get worse.

Upstream and firm energy price reforms

Policy needs to recognise concerns over possible carbon leakage and this can reasonably result in some firms facing lower effective costs than others. But many of the inconsistencies in prices across firms otherwise appear to be more the result of policy inconsistencies and the compounding effect of multiple policy instruments than the result of any clear objective. Many firms currently face an array of different rules, inclusion criteria and reporting requirements. Attempts to reform energy pricing for firms should focus on simplifying an increasingly complicated policy landscape and eliminating inadvertent inconsistencies. This would result in more consistent carbon prices and reduce administrative complexity.

To bring this about we suggest merging the Carbon Reduction Commitment Energy Efficiency Scheme (CRC), Climate Change Levy (CCL) and Climate Change Agreement (CCA) into one single instrument, in the form of a CCL-style carbon tax. This would be imposed at a uniform tax rate across firms and fuels.

There would of course be a reduction in the case of electricity to account for the upstream policies – the EU ETS, the Carbon Price Support Rate (CPSR) and market support for renewables – which in any case impose an effective carbon tax. If one were to aim for a total carbon price near to £65/tCO_2e by 2020 then the appropriate additional tax on electricity consumption may be very low.

Ideally, we would want international agreement leading to mutually acceptable (ideally uniform) carbon prices across jurisdictions. In practice we are a long way from that.

The EU ETS has not been effective and faces a number of challenges related to the inability to tighten the supply of emission allowances. The fact that the effective carbon price within the ETS has averaged just £3.50/tCO_2e so far in 2013 illustrates this problem. The EU ETS should be reformed, to introduce a price floor and ceiling in permit auctions and tighter ETS caps. This would help to reduce the risk for investment that arises from uncertainty about future policy.

In the absence of wider agreement it will also be necessary to ensure appropriately targeted additional support to firms at risk of carbon leakage. Identifying the firms in need of compensatory measures has been hard in practice, but there may be scope to target firms more accurately in future. Existing evidence suggests that the identification of companies at risk of carbon leakage so far has generally been too broad. Sweeping exemptions from policies such as the CCL and the EU ETS are costly and are likely to be poorly targeted. This is an issue that requires further study, and we recommend that the government should explore more accurate approaches to determine which sectors might be genuinely at risk of significant carbon leakage.

The objective of the reform options is not to raise additional revenue, but to make energy pricing more efficient, more consistent and cheaper administratively. Any additional revenues raised should therefore be recycled, with a focus on mitigating the effect on trade-exposed, energy-intensive industries (assuming that they can be identified correctly). The reforms are therefore intended to be tax neutral.

Household energy price reforms

Households face lower carbon prices than firms in large part because of the reduced VAT rate on residential energy consumption. Ignoring this VAT 'subsidy' the effective carbon tax on electricity is not far from the government's £59/tCO_2e. The effective carbon price on gas is much lower.

From an efficiency point of view, then, there is a strong case for imposing VAT at the full rate on domestic energy consumption, and an additional carbon tax on gas use – one of 0.8p/kWh would bring the carbon price into line with that on electricity. The first of these policies would raise around £5 billion and the second £3.3 billion in revenue.

The problem is of course that such policies are regressive. Poor households devote a larger portion of their spending to energy than richer households. Raising energy prices is not popular (as Chancellor Norman Lamont discovered in 1993 when he tried to impose VAT at the full rate on domestic energy consumption). And prices have been rising in recent years – by 96% since 2005 compared with Consumer Price Index (CPI) inflation of 23%.

So any move in this direction would need to be accompanied by a substantial compensation package. Note though that actual and planned policies, which have added 16% and will add 35% to the price of electricity, have not been accompanied by any such compensation package.

By generating over £8 billion in additional revenue, the policy changes we illustrate would allow a substantial compensation package for vulnerable households. The distributional effects of an example compensation package focused on increasing means-tested benefits are illustrated in Figure ES.2. At a cost of £7.2 billion this would leave households in the bottom third of the income and expenditure distributions better off on average, and leave relatively few net losers at the bottom of each distribution.

Figure ES.2: Average net relative impact of reform package, by expenditure decile

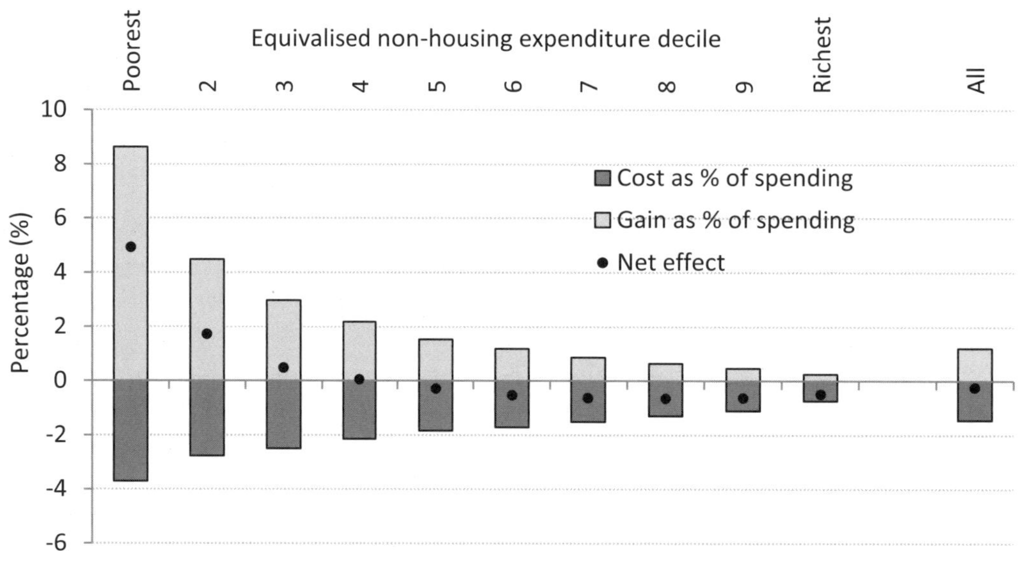

Source: Advani et al. (2013).

Such a reform would rationalise carbon pricing, be progressive across the majority of the distribution and still leave some surplus revenue for other expenditure priorities. Other compensation packages could include some income tax cuts and a less specific focus on the poorest. The key point is that a whole range of compensatory policies is possible and suggestions for reforming energy taxation should be seen in combination with such policies.

Of course, there would be both winners and losers from any such policy reform and the barriers to such a change would inevitably be considerable, particularly in light of the marked increase in energy prices in recent years. However, the principle that we could compensate most low-income losers clearly holds. Barriers to such a reform may also be reduced by introducing the policy in small steps over a number of years. This would avoid relatively large sudden increases, and therefore reduce the distributional consequences in any given year.

Crucially, it is important to be clear that the future context will be different. Effective carbon prices on electricity will rise substantially in the next few years in any case as a result of policies aimed at supporting renewable technologies. The extension of full rate VAT could therefore result in a future household carbon price significantly above the target price if other planned policies remained unchanged.

However, even without our proposed reforms there is a strong case for looking again at the funding of existing planned policies. As it is, the planned reforms imply carbon prices for business that significantly exceed the target price. The costs of this will still ultimately be borne by households. Furthermore, the widening gap in carbon prices faced by households and businesses only exacerbates the efficiency cost of carbon reduction policies. Hence there is a strong case for imposing VAT as we have suggested, with compensation, and exploring alternative ways of funding policies with additional aims. This ensures that achieving the desired carbon price is not dependent on the wholesale costs of energy, as it is when the VAT subsidy is used to influence the price, whilst also mitigating much of the distributional consequence of the price increases.

The efficiency gains from the reforms may be large, given the need to reduce emissions and evidence that the (relatively protected) household sector has previously lagged behind in this regard. The precise impact (on revenue and emissions) will depend on households' price elasticity of demand for energy. Previous estimates suggest this might be of the order of –0.3 in the short term and –0.8 in the long term. This would imply that the reforms could reduce household emissions by around 8 million tonnes of CO_2 per year in the short run, and more than 22 million tonnes a year in the long run, while still generating sufficient revenue to address distributional concerns.

The government also offers direct support for energy bills to some household groups through policies such as the Warm Home Discount. However, identifying precisely who is eligible is currently problematic and results in lower take-up of support among some household types. Take-up could be improved by sharing data between energy companies and government departments so as to reduce the administrative burden of delivering support. Legislation could be introduced to facilitate this. Automatic payment to eligible households would also reduce any social stigma attached to applying for such support.

There are also other potential reforms that could further increase the efficiency of bill support. Winter Fuel Payments are currently universal for all individuals over the female state pension age. However, the objectives of the policy are unclear, lying somewhere between transfers for energy bills and broader support for pensioners. Longer-term reforms may consider wrapping up similar bill support policies under consistent eligibility criteria that target all vulnerable households.

Introduction

Recent governments have been extremely active in policy related to energy use by households and businesses. Governments are pursuing a number of economic, social and environmental objectives related in particular to cheap, reliable supply, fuel poverty and decarbonisation. They also have fiscal objectives related to the energy sector. Several of these objectives have been formalised as explicit, often statutory targets. These include commitments on fuel poverty, the rolling five-year 'carbon budgets' aimed at the reduction of greenhouse gas (GHG) emissions and the EU Renewables Directive, which requires that 15% of all energy be renewably sourced by 2020.

Energy represents a significant share of household and business expenditures. According to the 2011 Living Costs and Food Survey, the average household spends 6% of their budget on energy. This budget share is considerably more for some poorer households, with an average budget share of over 15% for the 10% of lowest spending households. Policies aimed at reducing emissions by increasing energy prices have distributional impacts by their very nature. Targets related to fuel poverty therefore often conflict with other environmental objectives.

Policy design has significant implications for the cost of meeting these aims. It is crucial to understand the current set of policies that are in place and the implications that are associated with achieving these objectives. Thinking about ways in which these policies could be improved is imperative in order to design more effective future policy.

This report provides a broad overview of energy and climate change policies in the UK. We focus on policies related to energy used by households, businesses and the public sector for the purposes of heating, lighting, cooking and powering appliances. This accounted for around 60% of end-user GHG emissions in the UK in 2011.[1] Energy use associated with transport and waste is not included.

The report is organised as follows. Chapter 1 sets out the key economic principles that are relevant to energy use policies. We first focus on the negative external costs associated with carbon and GHG emissions that result from energy use, and discuss the importance of uniform carbon pricing in order to address these costs. We then consider reasons why we may wish for carbon prices to vary, and the various policy instruments that can be used to set prices. We also examine other non-carbon-based market failures associated with energy use, and how policy may address these issues.

Chapter 2 outlines the set of targets facing the UK government which are likely to influence decisions about energy use policies. We consider targets that relate to

[1] The authors' calculations using DECC official statistics are available at https://www.gov.uk/government/publications/final-uk-emissions-estimates.

three different issues: those related to GHG emissions, including carbon budgets and renewable energy targets; equity-based targets that aim to reduce fuel poverty; and a fiscal target to increase the share of revenues sourced from green taxes. We provide evidence on progress towards achieving these aims and offer commentary on the principles or practicalities of each target.

Chapter 3 describes the current set of policies related to energy use in the UK. We provide a brief overview of each measure, outlining the structure and the main objective of each policy. We describe the revenues and expenditures associated with each policy.

Chapters 4 and 5 provide commentary on current policies in three broad groups: Chapter 4 addresses those which affect all users through upstream energy prices[2] and those which affect downstream energy used by firms and industry; and Chapter 5 describes those which affect downstream energy used by households.

Chapter 6 assesses the implications of policy for variation in effective carbon prices on energy use. We calculate carbon prices across different fuels and for different end users. We present prices under the current policy landscape and predicted prices under the set of policies in place in 2020.

Chapters 7 and 8 consider options for upstream and firm policy reforms. We make recommendations that include both shorter term practical changes that could be made within the current policy environment, and longer term reforms. These recommendations are organised into the three categories described in Chapters 5 and 6, and are based on the discussion in the rest of this report, along with the findings of the work in two companion papers (Advani et al., 2013; and Bassi, Dechezleprêtre and Fankhauser, 2013). The final chapter concludes.

[2] The energy debate often distinguishes between upstream (the production and refining of fuels), midstream (the conversion of these fuels into electricity) and downstream (consumption by end users). For simplicity we label all forms of energy production, including power generation, 'upstream'.

1. Key economic principles for energy use policies

Arun Advani, Alex Bowen, Paul Johnson, Andrew Leicester and George Stoye

This chapter outlines some of the economic principles that will underlie any consideration of the current policy landscape towards energy use. There are several reasons why the price of energy should not be determined purely by the market. Many of them concern environmental costs (or 'externalities') related to air pollution and GHG emissions, but there are also equity issues related to fuel poverty and market failures related to technology innovation and energy efficiency.

We begin with a focus on the negative external costs associated with carbon and other GHGs emitted by the use of most forms of energy. We discuss the nature of the externality and issues in how GHGs should be priced – what price should be set, to what extent should prices vary across types and users of energy and what instruments could be used to set prices? We then look at other issues: what other market failures are important to consider and how might they be addressed by policy? This includes behavioural failures by energy users and issues in innovation related to energy use.

It is worth stressing that, ideally, policy would be set internationally. GHG emissions affect the world wherever they happen to be emitted. One would want carbon prices and policies to converge across countries over time. This lack of convergence is one reason to be concerned about 'carbon leakage' – the possibility that energy-intensive firms facing a high carbon price in the UK might move (some) production to a low-tax jurisdiction to take advantage of lower carbon prices elsewhere. This issue is discussed in greater detail in this chapter.

1.1 Carbon pricing: greenhouse gas emissions from energy use

The use of non-renewable energy by an individual consumer or firm imposes external costs on wider society. Burning gas to heat a home, running appliances powered by electricity generated from non-renewable energy sources or using coal to produce iron emits carbon dioxide and other GHGs that contribute to climate change. The market price of energy will not reflect these externalities (costs imposed on others), which will, in the absence of intervention, lead to an over-consumption of non-renewable energy relative to the level that would be socially optimal.

The most commonly advocated solution is to impose a charge on non-renewable fuels such that the total price reflects the full social costs of their consumption. This 'internalises the externality', reducing energy use to optimal levels and leading to an overall improvement in social welfare. This approach is usually referred to as 'putting a price on carbon', though in principle it is not just carbon emissions from energy that matter for global warming, but all GHGs. The contribution to global warming of non-carbon GHGs is usually expressed relative to carbon (see Box 3.1), such that a 'carbon price' acts as a sufficient shorthand description of the policy response.

There are a number of important issues in thinking about carbon pricing to which we now turn. The most obvious question is how the appropriate rate at which a carbon price is set should be determined.

The usual approach to pricing externalities, stemming back to Pigou (1920) and thus 'Pigouvian tax', argues that the price should reflect the external cost generated *at the margin*: for example, the extra costs of global warming resulting from burning an extra kWh of natural gas for cooking, or an extra tonne of coal for industrial processes. Of course, this is formidably difficult to estimate given uncertainties about the costs of climate change and the fact that these costs are related to the stock of GHGs in the atmosphere in the future, not just the current flow of emissions.

An alternative is to estimate the carbon price commensurate with achieving particular climate change targets. This is the approach used by government analysts to value carbon in policy appraisals, though of course uncertainties remain here, predominantly around the marginal costs of reducing emissions to the required levels. We draw on the government's carbon values in developing our recommendations for reform to household and business energy taxation, as discussed at the end of Chapter 6. A broader discussion of issues around setting carbon prices can be found in Department of Energy and Climate Change (2009) and Section 5.5.2 of Fullerton, Leicester and Smith (2010).

1.2 The rationale for a consistent carbon price

Irrespective of the method used to value the external costs imposed by emissions, a key question is whether emissions are priced consistently. In terms of climate change, the environmental damage caused by emissions depends only on the amount and type of each GHG emitted, not on who does the emitting and where.[3] As a result, a single emissions price is the most efficient outcome. If there are other externalities associated with using particular forms of energy then they too should be accounted for in any pricing regime.

A price for emissions would give energy consumers incentives to reduce emissions until the point where the cost of further reducing emissions is greater than the cost of paying to emit. If different users face different prices, then those facing a higher price would spend more per unit of abatement (emission reduction) than those facing a lower price. This creates an efficiency cost: the

[3] This is broadly true, but there are some exceptions. For example, non-carbon aviation GHG emissions may have local warming effects nearer to flight paths (IPCC, 1999).

same abatement could be achieved at lower cost by shifting abatement effort from the high-price to the low-price consumer.

Figure 1.1 illustrates the point. Imagine that the government has a target to reduce carbon emissions by a given amount, given by the length of the horizontal axis. Carbon is generated by two groups of energy users: firms and households. The target is for total abatement: if firms reduce carbon emissions by one additional tonne, households can emit one tonne more. The amount of carbon abatement by firms is measured from left to right, and by households from right to left. The vertical axes give the abatement costs for firms (left-hand axis) and households (right-hand axis), with the marginal abatement costs for firms and households given by the curves MAC_F and MAC_H. In this example, firms have relatively lower marginal abatement costs for any given level of abatement (their marginal abatement cost curve slopes upwards more gently).

Setting a consistent carbon tax of T_0 would see firms optimally shouldering more of the total required abatement (since their costs of doing so are lower) and generate the necessary carbon reduction at least cost. However, governments may be concerned that carbon taxes on households have negative distributional consequences (discussed below), and so prefer to reduce the household tax to T_H. To achieve the necessary abatement this requires increasing the tax on firms to T_F. Although this generates the same total carbon abatement, it is at a higher cost since even more of the burden is now placed on firms rather than households. The additional cost relative to the consistent price is given by the shaded area. This represents the efficiency cost of having an inconsistent price on carbon across sectors.

Figure 1.1: The efficiency cost of carbon price variation

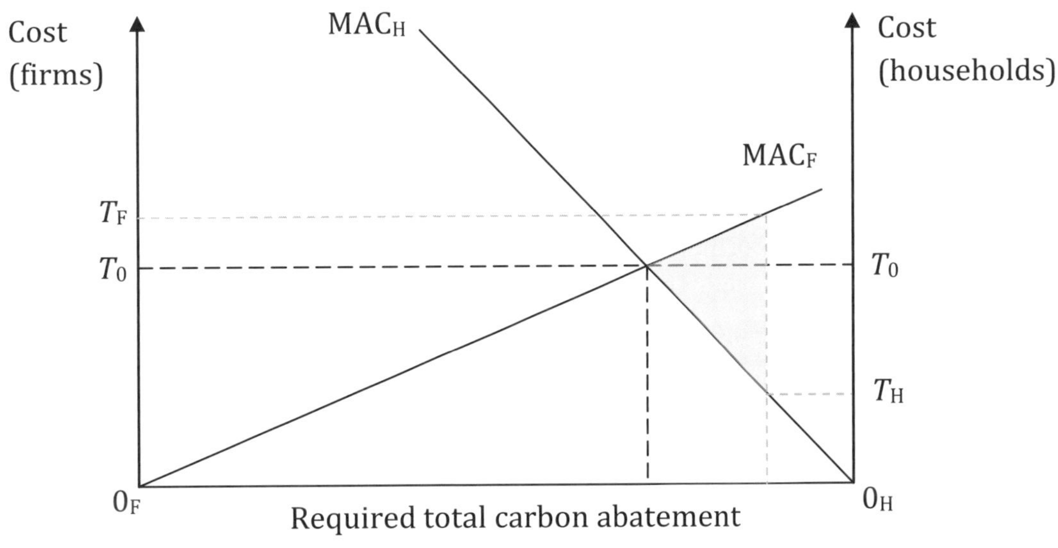

In Chapter 6, we estimate how carbon prices vary across end users and different fuels on the basis of current UK policies. In principle of course, setting consistent prices for carbon is not just an issue within the UK but a global issue, given that the costs of carbon emissions are not location-specific. This suggests the longer term objective should be to reach global agreement on carbon pricing. Many of the issues raised elsewhere in this report (including e.g. the threat of 'carbon leakage' we discuss in the next subsection) would be addressed by a truly global system of carbon pricing.

There is a clear efficiency case for setting a consistent carbon price in a single period of time. The same argument can be extended to setting a path for carbon prices over time. Suppose that Figure 1.1 were viewed not as two different groups of energy users, but as two different time periods (this year and next, say), with a required total abatement over both years. We might expect the marginal cost of abatement to fall next year – new abatement technology becomes relatively cheaper, for example. However, minimising the cost of total abatement over the period still requires the same carbon price in each year. More abatement will be done next year when it costs less to do so: the marginal cost of abatement in each period will then be equalised, meaning the total abatement is done at least cost. The subtle difference is that we want to account for the fact that future costs will be discounted relative to today. Thus, carbon price should rise in line with discount rates to ensure that the (appropriately discounted) incentives to abate at the margin are the same in each year. Real interest rates, typically around 3 to 5% per year, are often used (Bowen, 2011).

Even with an agreed level of carbon price and trajectory for how carbon prices change over time, there is an important point that learning and new information could lead optimal carbon prices to change. For example, in Figure 1.1 it may be thought that firms have much higher marginal abatement costs than they actually do. This would lead to a higher carbon price than T_0 being set, which actually leads to excessive amounts of abatement relative to the desired amount. Or the total desired abatement should change in the light of new scientific evidence. For example, if abatement levels are set to be consistent with a target level of future global warming, if it emerges that higher concentrations of atmospheric carbon would be consistent with that level of warming then less abatement, and thus lower carbon prices, would be appropriate. Equally, if the evidence suggested that a lower degree of overall warming ought to be the target then a higher carbon price would be needed.

Of course the information requirements to set the right prices are extraordinarily high, and new evidence is emerging all the time. Given issues around policy certainty (which we discuss below in the context of the long-term nature of many of the relevant decisions made about energy production and use), changing carbon price levels and trajectories constantly in the light of this evidence would be unhelpful. Ultimately there will be political and value judgements alongside the scientific and economic evidence that is used to inform the view of the right carbon price. However, it would clearly be helpful to review new evidence periodically to ensure that policy is not too divorced from something which looks appropriate given our current knowledge of the complicated issues involved.

1.3 When might we want carbon prices to vary?

Although a single carbon price can correct the externality caused by carbon emissions, there are a number of other issues in the market for energy that might rationalise a departure from a consistent price across all energy users. These include: equity considerations; carbon leakage; non-carbon externalities; issues of hassle costs and salience of carbon prices; myopic decision-making; principal–agent problems; and dynamic efficiencies in research and development. If targeted policies can be designed to tackle these issues directly, then the carbon price can be used to deal with only the externality directly caused by carbon emission. Where this is not possible, it might be necessary to vary the carbon price to best achieve the multiple objectives.

In this subsection we focus on the issues of equity and leakage. These concerns have been the main reasons for existing deviations from a consistent carbon price. In Chapter 1.6 we discuss some of the other topics.

Equity

Energy is a classic economic necessity: it makes up a larger share of total spending for poorer households than it does for richer ones. As a result, a price on carbon applied to all energy use equally would be regressive. Figure 1.2 shows that in Britain in 2011 the poorest tenth of households (measured by their total expenditure) devoted, on average, 15.8% of their budget to energy (electricity, gas and other fuels in the home) compared with 3.3% for the richest tenth.[4]

Since in general it is assumed that (all else being equal) a more equitable distribution of living standards is socially preferable to a less equitable distribution, there may be a willingness to set a lower carbon price for households relative to business energy users, trading off the efficiency loss outlined above against the equity benefits. Policymakers may also want to target lower carbon prices on poorer energy users, though 'means-testing' support in this way can lead to problems of identifying people eligible for support, administrative complexity, take-up and stigma issues, and of reducing the incentives for those affected to work if support is conditional on income or employment status.

[4] Total spending is often preferred to total income as a measure of household welfare since low income may not reflect low living standards if people are able to borrow or draw on savings, or if very low incomes are poorly measured in household surveys. Expenditure may better reflect longer term living standards than a snapshot of current income.

Figure 1.2: Average energy budget shares, by expenditure decile and fuel type, 2011

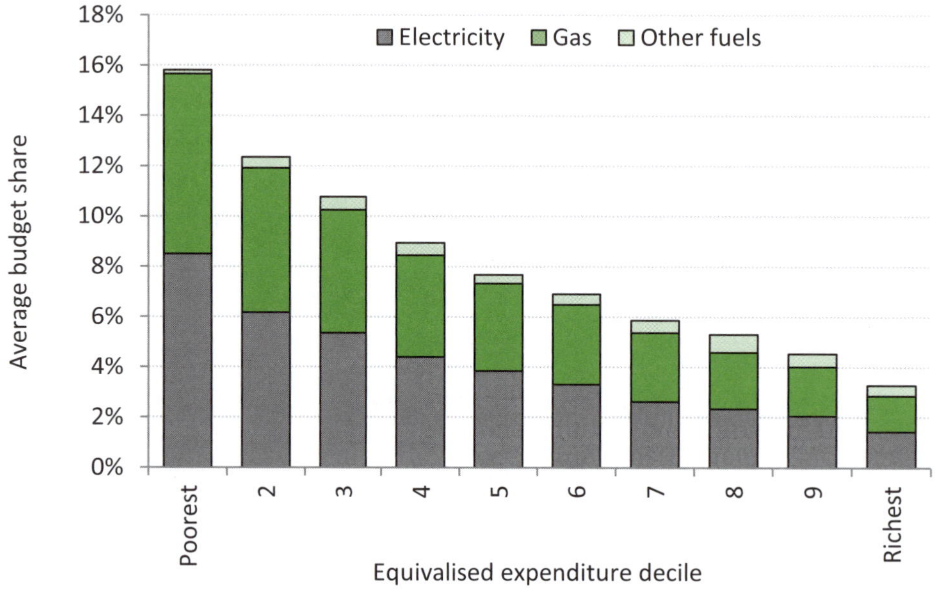

Source: Authors' calculations from Living Costs and Food Survey data. Note: excludes Northern Ireland.

An alternative approach would be to set the same carbon price across households and other energy users but then try to compensate the negative distributional consequences in some way, using existing mechanisms through the tax and benefit system or through creating some new income transfer policy.

This latter approach has the advantage of setting the right price incentives for household energy use, at least so long as the redistribution policy itself is not contingent on energy use. However, it may be difficult to redistribute effectively through other mechanisms. Tax and benefit policies usually vary along a number of dimensions such as income or household composition, and energy use varies substantially even for households with similar demographic characteristics. Evidence from the Department of Energy and Climate Change (DECC), for example, drawing on the National Energy Efficiency Data-Framework (NEED) database for Britain in 2010 suggests that among households with an income of less than £10,000 per year annual gas consumption ranged from 2,500kWh at the 5th percentile to 20,300kWh at the 90th percentile.[5]

This suggests that it would be extremely difficult to compensate *all* poorer households fully for a significant increase in the carbon price faced by households. However, it seems possible to compensate most poorer households, as we illustrate in Chapter 8, where we consider a package which increases household carbon prices with some of the revenue spent on increases in means-

[5] Data are available from https://www.gov.uk/government/uploads/system/uploads/attachment_data/file/65983/6949-national-energy-efficiency-data-consumption-distri.xls.

tested benefits.[6] It may also be possible to address distributional concerns in other ways (such as supporting energy efficiency). The trade-off between the efficiency gains from consistent carbon pricing and concerns about the distributional impact for those poorer households who are most difficult to help is of course the main issue for policy.

There are likely to be adverse distributional consequences from any policies – taxes, trading schemes or direct regulation – that increase the cost of energy use. This is an important point that often appears to be neglected in policy design: policies towards household energy use have tended to favour regulation over direct pricing, but these regulations do impose effective carbon prices on households (see Chapter 5). If policies do not raise revenue for the government (e.g. regulations and efficiency standards, carbon trading with free allowances) then the costs of compensating poorer households have to be borne from increased taxation or reductions in other spending (which in turn have their own distributional consequences). In general, then, the desire to compensate poorer households points to a preference for policies that generate revenues which can be used in whole or in part to fund the compensation policy.

More generally, at least some of the final cost of policies imposed outside the household sector is also likely to fall on consumers. Taxes on business energy use, for example, will be reflected in some combination of higher prices, reduced employment and lower profits. Each of these outcomes will have its own distributional consequences, but precisely what these are is not clearly understood, and thus they will be much more difficult to ameliorate properly through other mechanisms than when policies affect final consumers directly. Again, this important point is often overlooked in analysing policy proposals.

Carbon leakage

Distributional concerns about the impact of policies on poor households may be one reason why carbon taxes might vary if the first-best solution of appropriate redistributive payments is impractical. Another relates to concerns about the impact of policies on particular firms, particularly those whose operations can easily relocate abroad.

A given carbon tax will have a relatively larger impact on businesses which use more energy relative to other inputs to production, that is those that are more energy-intensive. In the absence of a global carbon price, if firms can choose where to locate, increases in carbon prices may cause some production to move to a low-tax jurisdiction. From an environmental perspective, total emissions do not fall (indeed, if the new country has generally laxer environmental standards, emissions may rise). From an economic perspective no revenue is raised and domestic firms suffer an undue loss in competitiveness. This threat of 'carbon leakage' therefore poses an additional constraint on setting carbon prices and

[6] Further details are given in Advani et al. (2013).

may rationalise lower rates being set for firms where the risk of carbon leakage is high.

However, identifying the firms or sectors where carbon leakage is a serious concern is not straightforward. In practice, policymakers have tended to rely on measures such as energy costs relative to turnover and trade intensity. Measures based on averages may not be good indicators of energy and trade intensity at the margin, i.e. which firms are most likely to 'leak' following a small increase in energy costs, and drawing an in/out distinction as to whether a firm is at risk of leaking is clearly somewhat arbitrary. The risk of carbon leakage also depends on other determinants of business location and expectations about future climate policy regimes elsewhere.

In 2009 the European Commission classified 164 sectors as 'at risk of carbon leakage' under the European Union Emissions Trading System (EU ETS). Subsequent research suggests that the risk of carbon leakage is mostly confined to a much smaller set of sectors, namely steel, cement, some basic chemicals, aluminium, pulp and paper and refineries (see, for example, De Bruyn, Nelissen and Koopman, 2013; Carbon Trust, 2010; Dröge and Cooper, 2010). De Bruyn et al. (2013) suggests that more realistic assumptions regarding ETS carbon prices, including 2020 prices below the €30/tCO_2 assumed by the European Commission, and supply and trade conditions, including the participation of new countries within the scheme, should be made. Under these assumptions, the proportion of sectors deemed at risk is reduced from 60% of sectors (accounting for 95% of industrial emissions) to 33% (10% of emissions).

In the absence of global carbon pricing, the ideal solution to this problem would be to set the same domestic carbon price across all firms, but to compensate those at risk of carbon leakage in some way unrelated to energy use. For example, some firms could receive rebates or other transfers related to their size or the average energy use across other firms in the same business sector. As with households, the question is whether the compensation could be delivered to the right firms and would be sufficient for those with the most severe risk of carbon leakage. Another option would be to reduce non-carbon taxes (such as corporation taxes) so that overall business costs are not increased, though again doing this in a way that targets firms most at risk of carbon leakage may be difficult. The Carbon Trust (2010) argues that solutions should be tailored to the needs of specific sectors, rather than generalised across industry.

Another solution would be to levy a tax on goods and services imported from countries without similar environmental standards or carbon tax rates to domestic ones, where the tax depends on the carbon embedded in the imports. Exported goods and services would have some of the carbon tax rebated, so that the implicit carbon price is more similar to the country to which the products are being sold. Such 'border adjustments' would level the playing field between domestic and foreign producers when selling in the UK, and ensure that the right price signals at the margin are given to firms. However, this comes at the cost of muting the impact of a carbon tax on emissions, since only output sold domestically is fully covered by the tax.

In practice it could be prohibitively costly to measure accurately the embedded carbon tax already paid on imported goods from different countries, and there is controversy about the legitimacy of such measures under international trade (World Trade Organisation) rules (Horn and Mavroidis, 2011). There is also a risk that any border adjustment would be captured as a means towards more general protectionism rather than being strictly applied as an environmental leveller.

Although they have economic merit (Helm, Hepburn and Ruta, 2012), it therefore seems impractical in the short term to use border adjustments to correct for the lack of any global agreement on environmental standards or a worldwide carbon price. In the longer term, consideration might still be given to how a system could be made feasible at relatively low cost without breaching trade rules (Ismer and Neuhoff, 2007), both as a backstop to continued lack of progress at an international level and perhaps as a way of spurring such agreement by providing a credible alternative proposal. In a carbon constrained world, the support required to prevent carbon leakage would decline over time as more stringent carbon policies were enforced in other countries (Committee on Climate Change, 2013a).

1.4 Instruments for setting carbon prices

A number of different policies could be used to set a price on carbon. The most obvious is a tax, which sets such a price directly. In principle, the ideal tax would be levied on GHG emissions themselves, as these are the source of the external damage. In practice, direct measurement of emissions is difficult or impossible, and so taxes tend to be levied on fuels which are burned to create energy. Since there is a very close relationship between the use of fuel and the ultimate emission of carbon, this is generally not an issue in the energy sector, although the advent of technologies such as carbon capture and storage could make that relationship more complex.

Another issue, in particular for electricity, is whether taxes should be levied at the point of use ('downstream') or of generation ('upstream'). A tax on electricity use could be set according to estimates of the GHG emissions associated with generating and supplying electricity, or levied on the fuels used as inputs to the generation based on the emissions content of the fuels.

Assuming that taxes on electricity generation are ultimately largely incident on consumers (given that the demand for energy is relatively insensitive to price, see Box 1.1) then it may not matter too much in practice at which stage of production or consumption the tax is levied. Electricity prices are largely determined by the 'marginal generator' – the type of plant which would generate the supply to meet an additional unit of demand. Thus, if a downstream tax were set at a rate commensurate with the carbon content of the marginal upstream generator, the final effect on price would be similar.[7] If instead a downstream tax was set on the basis of the *average* carbon content of all electricity, the two could be somewhat different.[8] There would also obviously be practical differences in the way upstream and downstream taxes would be administered. Section 5.5.3 of Fullerton, Leicester and Smith (2010) has a deeper discussion of the issues around levying taxes on energy generation ('upstream') or use ('downstream'). This report considers taxes on energy at all points in the generation, supply and use process.

Taxes are not the only policy which could be used to price emissions. Other economic instruments, such as emissions trading schemes, also impose effective prices. The government sets a cap on the level of emissions,[9] allocates permits to energy users (either for free or through some auctioning process) where permits represent a level of allowable emissions, and then allows participants to trade permits with one another. By creating a centralised market on which they may be traded, a single price for emissions occurs. Put simply, where a tax sets a price for emissions and then allows the market to determine the impact on the quantity, a trading scheme sets a quantity and allows the market to determine the price. Whilst, under idealised conditions, taxes and trading schemes would have the same effect on emissions, when there is uncertainty about the costs and benefits of emissions reduction (as there surely is in reality) the two can have quite different effects and costs (Weitzman, 1974; Pizer, 1999).

Other interventions also impose *de facto* prices on emissions. Prohibiting the use of a certain type of fuel represents an implicit price that is infinitely high for the emissions associated with using that fuel. Fining firms who use certain emissions-intensive processes imposes prices related to the level of the fine. More subtly, regulatory policies which raise marginal costs of energy generation or supply can also be seen as setting effective carbon prices: the effect is essentially the same as a tax that has the same marginal cost impact, though it is often much more difficult to estimate the marginal cost impact of regulations than the marginal cost impact of direct pricing measures such as taxes.

1.5 The importance of revenue raising

Another reason why we may want to impose a tax on energy use, or implement another revenue-raising policy such as an emissions trading scheme with

[7] Although the aim of policy in this area is to set carbon prices in order to generate the right incentives, the structure of policy will have implications for economic rents earned by inframarginal generators, the revenue received by government and the ability to incentivise particular forms of generation. Upstream taxes provide more flexibility in determining the allocation of revenue and providing incentives to particular forms of generation.

[8] As discussed more in Chapter 6 and Appendix B, the marginal carbon content of UK electricity, which is largely assumed to derive from gas-fired generation, is somewhat less than the average carbon content, which is pulled upwards by more polluting coal-fired generation.

[9] Though as noted above, since 'emissions' are not directly observed, they are estimated on the basis of fuels used which are assumed to convert to a certain level of emissions depending on the type of fuel.

auctioned permits, is to raise revenue. We discuss the amount of money raised from taxes related to energy use in Chapter 3.

A key issue is whether a tax or equivalent policy on energy use is a relatively more efficient way of raising revenue than other taxes. Early economic theory explored whether taxes on different goods and services should vary on efficiency grounds. Ramsey (1927) argued that in order to minimise distortions to consumption behaviour taxes should be higher on goods whose demand is price-inelastic (the 'inverse elasticity' rule). Given that the demand for energy is relatively inelastic, certainly in the shorter term (see Box 1.1), this might rationalise relatively high taxes on energy.

However, this argument is based on a very simplified world, in which taxes on consumption are the only source of revenue and there are no other market failures. When taxes can be levied on income and wealth directly, the rationale for imposing higher or lower taxes on particular forms of consumption may be limited to cases where there are external costs or benefits that we wish to correct, or where certain expenditures are strongly related to work incentives (Atkinson and Stiglitz, 1976).[10] As described above, there is clearly an external cost rationale for taxing energy. The relationship between the demand for energy and for leisure is less clear since energy consumption is associated both with work and with leisure. Trying to tax 'work-related' energy use distinctly from other energy use seems formidably complicated.

A double dividend?

One aspect of the debate around 'green' tax reforms which has received considerable attention is the 'double dividend' idea. Taxes on goods such as non-renewable energy, which have external costs, not only correct the market failure (the first 'dividend') but also raise revenue, which can be used to reduce other taxes, such as income tax (the second 'dividend'). The second benefit arises because of the distortions imposed by taxes on decisions about how much to work; these distortions are ameliorated when other tax rates are cut using the revenues from green taxes.

There is considerable debate over whether such a double dividend arises in practice. A full discussion can be found in Fullerton, Leicester and Smith (2010). The basic argument (Bovenberg and de Mooij, 1994) is that taxes on energy lead to increases in prices, which reduces the real purchasing power of wages and so reduces work incentives in much the same way that taxes on earned income do. The net effect of an energy tax and cut in direct income tax on work incentives is therefore ambiguous.

It is also critical to note that the 'double dividend' is normally argued not in terms of growth or employment, but rather in terms of economic welfare. If green taxes are increased far above rates that might be rationalised by the relevant

[10] An extensive discussion of these issues can be found in Chapter 6 of Mirrlees et al. (2011).

environmental externalities then from a welfare perspective the reform will have negative effects. Some aspects of the modelled reform would seem to fall foul of this: for example, fuel duty rates would more than double. Whilst there is uncertainty over optimal rates of fuel duty in the UK, and perhaps some scope to raise them, it seems highly unlikely that the optimal rate is more than double current levels (Johnson, Leicester and Stoye, 2012).

Whether or not there is a double dividend, it is also important to bear in mind that the *possibility* only arises if revenues are raised from the energy policies implemented. Policies such as regulation or freely allocated trading permits could, like taxes or auctioned permits, yield an environmental dividend. But they will not raise revenue which could substitute for that raised from other distortionary taxes.

Other transfers

Taxes or other revenue-raising instruments are a form of transfer (from energy users to the government) important to energy use policies. However, there are a number of other important transfers to consider.

One is a flow in the opposite direction: flows from government to energy users which effectively subsidise energy use for some consumers. This includes energy tax reductions or exemptions for certain groups of consumer, known as 'tax expenditures'. We discuss this in more depth in Chapter 3, but note here that in fact the size of such tax expenditures in the UK appears to be substantially larger than the revenue raised from taxes or similar instruments on energy use.

Another important set of transfers can come from policies which raise energy costs at the margin (and so raise prices), leading to windfall gains to non-marginal producers. If the policy does not recoup this gain in the form of revenue (for example, handing out free permits in an emissions trading scheme), it will instead by captured by the inframarginal generators. There is evidence, which again we detail further in Chapter 4, to suggest that such transfers have also been very large. This again suggests the importance of capturing the gains in the form of increased revenues where possible, rather than allowing the windfall to accrue to producers.

1.6 Other market failures

Aside from the carbon-related external costs associated with energy use, there are a number of other market failures which would justify policy responses. From the outset, though, it is critical to recognise that none of these issues justify varying the level of *carbon* prices across consumers or fuels, but rather suggest other, specific forms of intervention as we discuss.

Non-carbon externalities associated with energy use

The external costs of energy use are not confined to climate change costs related to emissions of carbon and other GHGs. Burning various fuels may release local air pollutants such as particulates which affect the health of those living near to power stations or heavy industries. There are disamenity costs for those living near generation and processing plants for all sources of energy, whether fossil fuel or renewable. Costs may also be imposed on other natural resources, e.g. water pollution or habitat destruction. There are also costs to do with the risk of accidents, not all of which may be properly covered through insurance markets available for those who may be affected.

Another externality could arise from an idea of 'energy security'. If energy resources are primarily obtained from countries or states thought to impose some sort of security risk or threat, then increases in energy use could impose additional costs in terms of increases in these risks or extra expenditures designed to ameliorate them (Gillingham and Sweeney, 2010). It is worth noting that the concept of 'energy security' appears to have several different definitions (Cherp and Jewell, 2011), including the risk of fossil fuel supply shortages in the long term, which may rationalise moves towards renewable energy, or the risk of short-term supply constraints as energy infrastructure is replaced. However, neither of these definitions would lend itself to an externality interpretation, since presumably the internal costs of energy would adjust to reflect particular short- and long-term imbalances between supply and demand.

Other externalities associated with energy use could rationalise intervention. If these other externalities vary across energy type, user or time, the tax rate could, if practical, also vary along similar dimensions. In principle the tax rate should reflect the size of the *marginal* externality, rather than the average.[11] However, estimating the appropriate marginal externality is extremely difficult. It may not relate directly to energy use, breaking the correspondence between a tax on energy and a tax on the externality that existed in the GHG case. For example, air pollution depends on atmospheric reactions, and the landscape disamenity from onshore wind turbines is not linked directly to energy generated.

Consumer responses to energy price signals

Standard economic analysis assumes that energy users (households and businesses) respond to price signals and optimise their consumption behaviour accordingly. As a result, policies which raise the price of energy relative to other goods and services (or other business inputs) will reduce energy demand as consumers adjust their purchasing behaviour. There is empirical evidence that energy consumers respond to higher prices by reducing their demand. Box 1.1 provides a brief summary of some of the literature.

[11] Bovenberg and Goulder (1996) note that when externality-correcting (Pigouvian) taxes are levied alongside taxes on income that distort labour supply, the optimal Pigouvian tax will be less than the marginal externality because of additional distortion to labour supply caused by the reduction in real wages when prices rise. This is related to the debate around double dividends from environmental taxes (see Section 2.5.2).

Box 1.1: Evidence on the relationship between energy prices and energy demand

A number of studies have estimated the price elasticity of demand for energy use (that is, how the demand for energy changes in response to a small change in relative energy prices). A survey of studies looking at residential energy is given in Espey and Espey (2004). In the short run a 1% rise in domestic electricity prices reduces demand by around 0.35%, whereas in the long run demand falls by 0.85%. The fact that demand is more inelastic in the short run is not surprising. As energy prices rise, the number of ways in which consumers can adjust in the long-term grows (buying more efficient appliances or fewer appliances, installing energy-efficiency measures, changing lifestyle behaviours). In the short term, taking all these factors as fixed, reductions in demand can only arise essentially though less heating, lighting, cooking and running of appliances. Demand may also be less responsive to price in the short term if there is uncertainty about whether a price change is likely to persist.

However, even in the long term, allowing for these other behavioural responses, electricity price elasticities are still less than 1, suggesting that demand is still not all that responsive to price on average.

Gillingham, Newell and Palmer (2009) summarise other evidence on short- and long-run elasticities across households and firms for electricity, gas and heating oils. It is hard to draw many firm conclusions given the range of estimates about whether electricity demand is more or less price-elastic than gas or oil demand, or whether firm demand is more price responsive than household demand. Nevertheless, the finding that demand is more elastic in the long run than the short run appears to be a robust result, as is the result that short-term elasticities are substantially less than 1, whereas long-run elasticities are closer to, or sometimes exceed, 1.

Behavioural economics has suggested a number of wider influences of consumer behaviour in energy markets which nuance the standard economic view. Here we offer a brief overview. Detailed discussions can be found in Ofgem (2011a), Pollitt and Shaorshadze (2011) and Wilson and Dowlatabadi (2007).

If price signals are not visible to consumers then they are unlikely to influence decision-making. Chetty, Looney and Kroft (2009) find that when sales taxes in the US are made more visible by including them on in-store price labels rather than being added at the till, consumption was reduced as if the price had increased even though the final (tax-inclusive) price was unchanged. In the context of energy use, consumers who pay by direct debit may not be aware of price changes, whereas those who receive a regular bill or use a prepayment system might be much more conscious of energy prices and so more responsive to them.

A key insight from standard models is that consumers should respond to *marginal* price incentives (for example, the cost of consuming an additional kWh of gas) rather than *average* price incentives (the total energy bill divided by total consumption). However, price signals may be complicated to understand, meaning that people fall back on rules of thumb (perhaps not responding to small price changes or responding to average prices rather than marginal prices if the former are easier to calculate). This is the idea of 'bounded rationality' (see, for example, Gigerenzer and Brighton, 2009). One implication is that firms may want to make price signals deliberately opaque (Ellison and Ellison, 2009). Even if consumers are able to see and understand price changes, their willingness to react to them may be limited if there are a large number of alternative consumption choices. This idea of 'choice overload' was popularised by the study of Iyengar and Lepper (2000), who showed that people were less willing to purchase jam when faced with lots of choices compared with fewer choices.

The way that energy in particular is priced may not be very transparent or understandable for some consumers: many tariffs include fixed components (a standing charge) with a variable unit price, or two different unit prices with a higher price for the first batch of energy used in a given period. Consumers might therefore respond to changes in their energy bill (which is the visible, understandable number, but is determined by the average cost of energy) rather than changes in energy prices at the margin. Friedman (2002) uses data on domestic gas consumption in the US and finds evidence that behaviour is better predicted by a model in which consumers respond to bills than by a model in which consumers respond to marginal prices.

There are also a large number of energy price plans and energy providers, meaning consumers face a large number of choices if they want to react to price signals by switching their provider. Ofgem (2011a), for example, estimate that consumers could pick from over 300 tariffs in 2011. This may act as a disincentive to switch, notwithstanding the number of comparison websites, which should make the process more straightforward if consumers have the necessary information about their energy use to get accurate comparisons.

Thus, policymakers need to consider not only the level but also the visibility ('salience') of price signals and consumers' ability to understand and respond to them. For example, energy companies could be required to price energy in a consistent way, or to limit the number of different tariff options they provide.[12] Of course limits of this kind impose costs on consumers who lose out from a pricing

---

[12] The Energy Bill currently before Parliament contains provisions to limit the number of tariffs that can be offered and to make tariffs across suppliers more easily comparable (see http://www.publications.parliament.uk/pa/bills/cbill/2013-2014/0004/en/2014004en.htm). An open question is what happens after the Smart Meter Roll-Out (see Section 4 and Appendix A) is completed. In principle, smart meters would allow more direct tailoring of tariffs to individual consumer energy-use patterns and allow consumers to make fully informed comparisons across providers using the information from their meter. This might conflict with legislation that limits tariff options.

strategy more suited to their particular consumption patterns: there is a trade-off between simplicity, in terms of making informed choices, and the benefits of variety, given the significant heterogeneity in consumer energy needs.

Firm responses to energy price signals

Issues around salience and complexity can apply to energy use decisions by firms as well as domestic consumers. Qualitative research by Carbon Trust and SPA Future Thinking (2012) finds that barriers to reducing electricity consumption in firms were related to firm characteristics such as size. Small firms reported that complexity and lack of expertise were barriers to reducing energy use; larger firms faced other problems in responding to energy price signals in terms of organisational structure. As firm size increases, decision-making structures become more complicated and there may be a disconnect between people responsible for decisions that affect energy use and people responsible for responding to policy incentives through taxation or other measures. In the latter case, then, there may be a role for policies which try to raise the profile of energy within firms' decision-making structures: publishing data on firms' energy use allowing for public comparisons, for example, might create reputational incentives, which could be more motivating than financial incentives in some cases. Regulation and auditing might also be needed if the response to price incentives is muddied by having many decision-makers within large firms.

Market failures in energy efficiency decisions

Energy used by households and firms is an example of a *derived demand*: in other words, people do not value the energy in its own right, but instead use energy to achieve other outcomes such as staying warm, heating water, cooking and powering appliances. These outcomes depend both on the amount of energy used and energy efficiency: more efficient buildings and appliances require less energy to heat or to run to achieve a given level of warmth or performance.

Typically there is a dynamic trade-off between energy efficiency and use. Energy efficiency measures in the home require upfront costs but reduce future heating bills. More efficient boilers, fridges, computers and televisions may be more expensive to buy but will be cheaper to run. For consumers, the standard economic model assumes they trade off the upfront costs of more efficient options against appropriately discounted (expected) future benefits. By increasing running costs relative to upfront costs, taxes on energy use could therefore encourage reduced consumption by making more energy-efficient choices optimal.

A study by McKinsey and Company (2012) for DECC suggested that electricity demand in the UK could be reduced by around 146 TWh (more than 35% against baseline forecasts) by 2030 if efficiency measures that were in principle *privately* optimal (under standard assumptions about how people discount future benefits against upfront costs) were taken up by households and businesses. This is an example of what has become known as the 'energy-efficiency gap' (Hirst and Brown, 1990; Jaffe and Stavins, 1994).

Understanding why energy users appear to be making less than optimal choices about efficiency is therefore crucial in terms of thinking about appropriate policy responses. A number of suggestions have been put forward; we summarise a few here, though a more detailed discussion can be found in Gillingham, Newell and Palmer (2009).

One possibility is that consumers weight upfront costs more heavily relative to future costs than is assumed in standard economic models. An early study by Hausman (1979) of the demand for air conditioners in the US found that consumers would have to discount future running costs very heavily – perhaps in the order of 15–25% per year – in order to explain their choice of cheaper, inefficient products. More recent models have suggested the idea of 'present bias', where discount rates vary over time: consumers have very high discount rates over the immediate future and then more moderate discount rates when they compare the immediate future with the more distant future.[13] One implication of this is that energy users can exhibit time inconsistent behaviour: they *plan* to install efficiency measures or to buy efficient appliances in the future when both costs and benefits are far distant, but then fail to do so once the purchase decision arrives and the benefits become heavily discounted relative to the costs.

From a policy perspective, present bias suggests that taxes on energy use (which alter future running costs) may need to be accompanied by other policies that alter upfront purchase costs such as subsidies to energy efficiency or purchase taxes.[14] There may also be a strong case for regulation which compels improvements in the efficiency of appliances or homes.[15]

The academic evidence on the importance of present bias is somewhat mixed. Greene (2010) reviews 25 studies looking at vehicle purchases. Only 12 of them find that efficiency is undervalued (in that consumers would not be willing to pay £1 more for a vehicle that cost £1 less to run in present discounted terms), as would be expected from time inconsistency. There appears to be little or no evidence on this issue relating directly to energy use in homes and businesses.

There could be other reasons why seemingly efficient investments are not taken up where the policy implications would be different. There may be hidden costs to installing energy efficiency measures beyond the upfront purchase prices: clearing out lofts and rooms to have insulation installed, for example. Factoring in

[13] DellaVigna (2009) summarises a body of evidence for present bias from a number of field experiments.

[14] Heutel (2011) models decisions over vehicle purchases under time inconsistency and suggests that optimal policy should include a tax to account for the externalities associated with fuel consumption and a policy to deal with time inconsistency, which could be a mandate or another price instrument which reflects the 'internal' costs consumers' time inconsistency imposes on themselves in the future when they face higher running costs.

[15] A number of such regulations are in place, including the Products Policy (see Chapter 4.1 and Appendix A).

these costs could make it optimal not to install measures even if there is no present bias, and helping to reduce them would be a sensible policy response. Caird, Roy and Herring (2008) find some evidence that hassle costs are a key barrier to installing loft insulation, for example, though this is based only on a small sample of individual respondents.

Another possibility is that people would like to invest in efficiency measures but lack the upfront capital to pay for them. In theory, people could borrow from capital markets assuming that anticipated future savings would be sufficient to repay the loan. However, energy efficiency savings are unlikely to represent robust collateral. Thus, credit constraints could be a market failure inhibiting take-up of cost-effective insulation or other efficiency investments relating to energy consumption. Policies which alleviate these constraints, either through subsidies or by providing a credit mechanism, would then be effective.

There is little evidence that factors which might be correlated with the presence of credit constraints, such as low income, low education or being out of work, are related to a lower probability of owning particular insulation measures such as loft insulation, cavity wall insulation or double glazing in England (Brechling and Smith, 1992; Leicester and Stoye, 2013). This might suggest that credit constraints are not the key barrier to take-up, at least for these measures, though of course they may be more important for expensive measures or packages of measures. The result could also reflect the impact of previous obligations on energy suppliers to improve efficiency, which have focused wholly or partly on poorer household groups (see Chapter 5.3 and Advani et al., 2013).

Finally, a broad class of market failures in energy efficiency relate to information. For example, it may be that consumer understanding of ongoing running costs relative to upfront purchase prices is relatively limited. Policies which provide information to consumers about running costs in a transparent and understandable way (allowing them to compare easily different options, such as efficiency classes of fridge) could be effective.

There may in some cases be a principal–agent problem in the market for energy efficiency if the person making the choice over efficiency investments (the principal, such as a landlord) is not the same as the person who pays the day-to-day bills to meet running costs (the agent, such as a tenant). Leicester and Stoye (2013) find that, all else being equal, those in the rented sector are around 11–14 percentage points less likely to have cavity wall or loft insulation than newly moved owner-occupiers. A landlord's willingness to install insulation measures in a home will depend on whether they can recoup the costs of the investment through higher rental prices. This requires that they can credibly signal to potential tenants that the measure is in place and so raise the rent compared with an otherwise equivalent property that has not had the measure.[16] The role of policy might be to regulate rental markets to require some independent agency

[16] And of course tenants understand why the rental price is higher (since heating costs would be lower). Here issues about valuing upfront versus ongoing costs also come into play.

that is trusted by tenants to verify the presence of measures, or indeed to mandate certain measures before a property can be let.[17]

Failures in the market for innovation

Energy efficiency will depend not just on consumer demand for efficient products but also firm investment in bringing such products to market. Similarly, new technologies to help capture emissions from electricity generated using fossil fuels or methods to reduce emissions from industrial processes will require innovation on the supply side.

In general, markets for innovation fail if the innovators cannot receive sufficient return to their investment: new technologies or processes can be adopted by others. This is not a problem specific to 'green' innovation, and a well-established patents system has developed to ensure that innovators can benefit from their investments. The wider positive spillovers from research and development would also justify subsidies to such investment: innovation can provide permanent productivity benefits to the innovator and other firms or wider society once a patent period expires.

There may be reasons to want to encourage investment in energy efficiency or low-emissions technology innovation that come from the interaction of the negative pollution externality with the positive social spillover effects of innovation and knowledge generation (Jaffe, Newell and Stavins, 2005). For example, if negative externalities from GHG emissions are not properly priced because of political or other constraints on setting carbon prices, then there may be larger social benefits from encouraging a switch from polluting to non-polluting technology. The reverse may also be true: if it is difficult to subsidise innovation directly then there may be an additional case for pricing the negative externality in terms of the incentives given to invest in socially beneficial innovation in clean technologies (Aghion et al., 2012).

In general, though, these would be second-best policy responses. Popp (2006) uses an economic and climate model and argues that in the absence of carbon pricing, R&D subsidies for green technology are by themselves fairly ineffective tools to deal with climate externalities because it is the pricing which encourages take-up of new technology. He also notes that, at least in the short-term when the supply of researchers and scientists is relatively fixed, additional subsidies to 'green' R&D are likely to crowd out R&D in other sectors. If the excess of social over private returns to R&D exceed that of green R&D, then this will reduce economic welfare. Optimal policy should try to provide the right carbon price incentives to deal with marginal external costs alongside subsidies to research and development to deal with the positive social spillovers, with subsidies targeted on sectors with the largest excess social returns (Acemoglu et al., 2012).

[17] Existing policies include Energy Performance Certificates, which give information to tenants and house purchasers about overall energy efficiency, and will also verify to new homeowners or tenants any Green Deal liabilities (see Chapter 4 and Appendix A) they would take on.

There is also an issue in terms of innovation and investment around the consistency of policy and providing certainty. Investment decisions are inherently long term, and will be influenced by expectations of how policy will change in the future, as well as other expected future changes. They are also hard to reverse once made if the future does not turn out as expected. Put another way, the decision to invest (in renewable energy, green R&D, energy efficiency, dwelling insulation and so on) will depend not only on the expected return but also the volatility of the return. Risk-averse investors may prefer not to invest if returns are very uncertain, or prefer to exercise the 'option value' of delaying investment to see how the future unfolds before making a committed decision (Hassett and Metcalfe, 1993). If policies are frequently revised in ways which raise the volatility of returns, or future policy announcements are not seen as credible, this could undermine incentives to invest.

1.7 Conclusions

The presence of environmental externalities and other market failures makes it is undesirable for the price of energy to be set purely by the market. However, market interventions need to be consistent in the way externalities are priced and take into account potential overlaps or trade-offs between different interventions and policy goals.

A single, consistent carbon price that provides the correct price incentives is the most economically efficient way to address the climate change externality and reduce carbon emissions. In a world where climate change mitigation is our only objective, policy should be designed so that carbon prices do not vary across fuels and users, or over time.

However, there are two important reasons why policy makers may wish to allow carbon prices to vary. First, equity concerns potentially play a key role. Consistent carbon prices for households across the expenditure distribution will impose a larger cost on poorer households, who typically spend a larger share of income on energy. This leads to the temptation to reduce the carbon price faced by poorer households, and results in a trade-off between efficiency and equity concerns. However, a more desirable approach would be to set a consistent carbon price across households, and compensate poorer households for the increased cost through other mechanisms, such as the existing tax and benefit system. This reduces the negative distributional effects while maintaining the correct price incentives for households energy use. This issue is explored in greater detail in Chapter 8 and in the companion paper.

Second, the threat of 'carbon leakage' may provide a rationale for lower carbon prices being set for firms where the risk of leakage is high. Energy-intensive firms with the ability to relocate may choose to move production to jurisdictions with lower carbon prices. Total emissions are not reduced (and may even increase if environmental standards in the low-tax jurisdictions are less stringent) while revenues are lost. In parallel to the first-best solution to household distributional concerns, the ideal solution would set a consistent carbon price across all firms and compensate those most at risk of carbon leakage in way that does not alter the incentives related to energy use. However, the practical identification of these sectors is not easy, and a detailed discussion is required over the best method of compensation. These issues are explored in detail in Chapter 5.1.

Other non-carbon market failures also need to be addressed, and that provides additional incentives to tax (or subsidise) energy use. However, they should be targeted separately through specific interventions and do not justify any variation in carbon prices.

2. Targets related to energy use policies

Arun Advani, Paul Johnson, Andrew Leicester and George Stoye

This chapter considers a number of current targets adopted by the UK government which are likely to influence decisions made about energy use policies. We consider targets related to carbon emissions (carbon budgets and renewable energy targets), equity-based targets to reduce fuel poverty and targets to raise a bigger share of revenues from green taxes, which would include taxes on energy use. In each case, we outline the nature of the target, progress in meeting it and offer some comments on the principles or practicalities of the particular target.

2.1 Reducing emissions

Carbon budgets

The Climate Change Act 2008 commits the UK government to reducing GHG emissions by 80% of their 1990 levels by 2050.[18] The government is obliged to set out legally binding 'carbon budgets': caps on total emissions over a five-year period, designed to be consistent with the 2050 target. Budgets are announced at least 12 years in advance. Advice is supplied by the independent Committee on Climate Change (CCC), also established by the Act, though budgets are approved by Parliament, and not the CCC.

Table 2.1: Current carbon budgets

Budget period	Total emissions ($MtCO_2e$)	Average annual ($MtCO_2e$)	Change from baseline
	1990 baseline	*774.3*	–
2008 to 2012	3,018	603.6	–22.0%
2013 to 2017	2,782	556.4	–28.1%
2018 to 2022	2,544	508.8	–34.3%
2023 to 2027	1,950	390.0	–49.6%
	2050 target	*154.9*	*–80.0%*

Source: DECC carbon budget information (https://www.gov.uk/government/policies/reducing-the-uk-s-greenhouse-gas-emissions-by-80-by-2050/supporting-pages/carbon-budgets) and emissions data (https://www.gov.uk/government/publications/final-uk-emissions-estimates), authors' calculations. Note: $MtCO_2e$ is million tonnes of carbon dioxide equivalent.

[18] The Act can be found at http://www.legislation.gov.uk/ukpga/2008/27.

Table 2.1 shows current carbon budgets, which have been legislated to cover four periods up to 2023 to 2027.[19] The budgets imply that by 2020, halfway between the 1990 baseline and the final target date, UK emissions will have fallen by around 34%. This suggests that the pace of emissions reduction is expected to accelerate in order to meet the overall 80% cut by 2050.

Figure 2.1 shows how emissions have changed between 1990 and 2011 and how future carbon budget targets compare with outcome emissions. The dark green line shows outcome emissions data between 1990 and 2011, the latest year for which final estimates are available. The grey dashed line shows a straight line path from the 1990 baseline emissions level to the long-run 2050 target to cut emissions by 80% from baseline. The red dotted lines show the average annual emissions permitted under the four carbon budget periods shown in Table 2.1.

Box 2.1: Emissions covered by carbon budgets

Carbon budgets govern the majority of GHG emissions produced in the UK. There are six gases covered: carbon dioxide (CO_2), methane, nitrous oxide and three gases (hydrofluorocarbons, perfluorocarbons and sulphur hexafluoride) collectively known as fluorinated compounds or F-gases. Emissions are normally expressed for each gas in CO_2-equivalent terms (CO_2e), based on estimates of the contributions of the emissions of each gas to global warming relative to CO_2 (over a given time horizon, usually 100 years). For example, it is estimated that a tonne of methane emissions contributes the equivalent of around 21 tonnes of carbon dioxide in terms of global warming, such that the appropriate tax rate on methane should be 21 times higher than that on carbon dioxide.

Carbon budgets currently exclude any emissions from international aviation and shipping which might be wholly or partly attributed to the UK. The CCC had previously recommended that an estimate of these emissions should be included in the measure,[a] but the government has deferred a decision of whether or not to do so pending a global deal to tackle international aviation and shipping emissions. However, the government has acknowledged that aviation and shipping emissions are included in the 2050 target,[b] and the current budget levels include some headroom that would allow for these emissions to be included in the future whilst still meeting an overall 80% reduction. Given that international aviation and shipping emissions are expected to fall by less than 80% by 2050, other sectors would need to cut by more in order to meet the overall objective. The most recent estimates, for 2011, suggest that international aviation and shipping generated 43 million tonnes of CO_2e emissions. If added to other sources, they would account for 7.2% of

[19] Though the fourth budget may be subject to review pending EU agreements on future carbon targets. See, for example, http://www.guardian.co.uk/environment/2011/oct/03/osborne-uk-carbon-emissions-europe.

emissions, a larger proportion than in any other year from 1990, when data are first available.[c]

For the purpose of carbon budget accounting, traded sector emissions are assumed to be exactly the UK's share of the EU-ETS cap. Territorial emissions are also measured and reported, although not accounted for carbon budgets.. In principle, this means the UK could make progress towards carbon budget targets by buying permits from abroad. This may be economically efficient if foreign emissions can verifiably be reduced more cheaply than is the case in the UK, though the CCC has recommended that budgets be met excluding net permit sales as far as possible.[d]

Carbon budget emissions targets are based on production-based measures of emissions produced in the UK. An alternative consumption-based measure of emissions shows a different trend in emissions over time.[e] Between 1993 and 2010, production-based GHG emissions fell by more than 19%, from 734 million tonnes of CO_2e to 592 million. The consumption-based measure *rose* by more than 5%, from 930 million tonnes of CO_2e to 981 million. Taking the change to 2007, prior to the onset of recession, the consumption-based measure rose by almost 25%, compared with a fall of 12.5% in the production-based measure. The different trends reflect the fact that the UK is a net importer of consumer goods, that these goods tend to have high embedded emissions and that they come from countries with higher emissions intensities than the UK. Arguably, a consumption-based measure is a more appropriate account of emissions generated by UK economic activity, though there is much more uncertainty about precisely how to measure it than the production-based definition. Continuing to estimate emissions on both measures would, however, be valuable.

[a]See http://www.theccc.org.uk/reports/international-aviation-a-shipping.
[b]See https://www.gov.uk/government/uploads/system/uploads/attachment_data/file/65686/7334-int-aviation-shipping-emissions-carb-budg.pdf
[c]https://www.gov.uk/government/uploads/system/uploads/attachment_data/file/73155/2011_final_uk_figures_data_tables.xls, Table 8.
[d]See http://www.theccc.org.uk/reports/fourth-carbon-budget.
[e]Data are available at http://www.defra.gov.uk/statistics/files/release-carbon-footprint-dec2012.pdf.

Figure 2.1: Trends in and targets for GHG emissions, 1990 to 2050

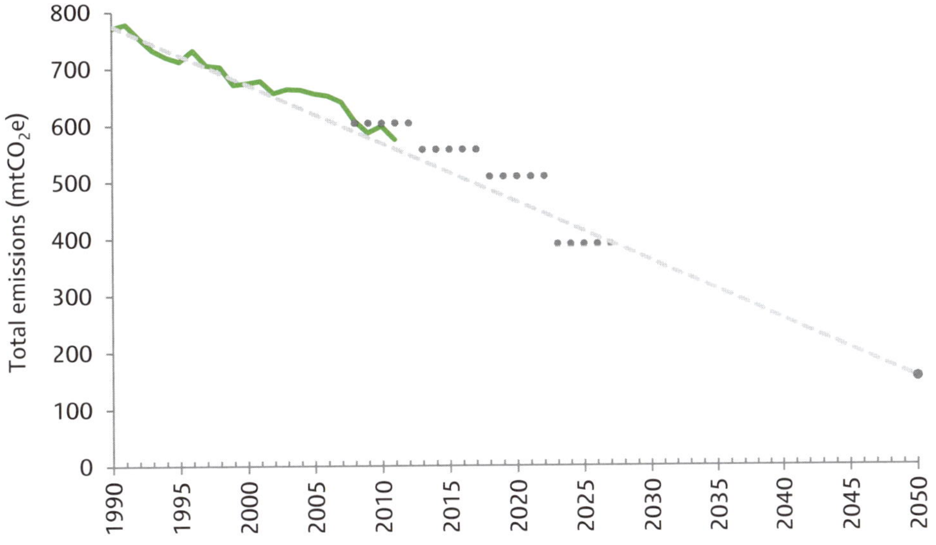

Source: Authors' calculations from DECC 2011 final emissions statistics (https://www.gov.uk/government/uploads/system/uploads/attachment_data/file/73155/2011_final_uk_figures_data_tables.xls, Table 9) and carbon budget data as in Table 2.1. Notes: grey dashed line shows linear path from baseline emissions to 2050 target. Red dotted lines show future carbon budgets in terms of average annual emissions allowed within each five-year budget period and the final 2050 target.

Emissions fell by around 26% between 1990 and 2011, from 771 million tonnes of CO_2 equivalent ($MtCO_2e$) to 574 million. This represents an average decline of 1.4% per year.

The decline can be seen in three broad phases. Between 1990 and 1999, emissions fell at around 1.5% per year on average, and were on the straight-line path towards the 2050 target (which of course had not at that point been set). This is the period often referred to as the 'dash for gas', during which much coal-fired electricity generation was replaced with cleaner gas-fired power (we return to fuels used for electricity generation below). From 1999 to 2007, emissions fell at a much slower pace, just 0.6% per year on average. Between 2007 and 2011, however, emissions fell rapidly, by 2.7% per year. This final reduction largely reflects the severe recession and loss of economic activity over the period.

By 2011 emissions were roughly in line with the straight-line path towards the 2050 target. As noted above, though, the carbon budgets allow for a slower initial pace of emissions reduction – the first, second and third budget periods to 2022 allow for emissions above the linear trend, falling below only in the fourth period from 2023. As a result, the UK appeared to be comfortably on course to meet its first carbon budget, which ended in 2012. Cumulative emissions between 2008 and 2011 were 2,365$MtCO_2e$. This meant emissions in 2012 could be 653$MtCO_2e$ without breaching the total five-year budget cap. This would represent a rise of

14% on 2011 emissions. Given the lack of growth in 2012, such a rise seems highly unlikely to occur.

Future carbon budgets become progressively more challenging to meet:

1. Being on course to meet the second budget in 2015 (the middle year of the five-year budget period) would require emissions to fall by around 0.8% per year on average from their 2011 values. This is slightly faster than the pace of emissions reduction in the early 2000s, though of course this was also a period of relatively strong economic growth.
2. Meeting the third budget in 2020 requires emissions to fall by 1.3% per year on average, roughly the same pace as the overall trend since 1990.
3. Meeting the fourth budget in 2025 requires emissions to fall by 2.7% per year on average. In other words, emissions would have to fall at the same pace as they have in the face of the severe economic downturn since 2007, even if growth returns to more normal levels in the medium term, as is currently forecast.

Of particular interest when thinking about policies that might affect energy use is how emissions vary across different sectors – e.g. the business sector, the domestic sector, the public sector, agriculture. If changes in emissions over time have varied across areas of economic activity, this might suggest either where previous policy has had the greatest impact (if changes can be attributed to policy reforms) or where future policy may need to focus.

There are two main ways in which we can break emissions down across sector: on a *source* basis and *end-user* basis. Source-based measures record direct emissions from different sources, whereas end-user measures additionally allocate the emissions from power generation to different users, which we might think of as indirect emissions resulting from the demand for electricity by different groups. When thinking about households, for example, direct emissions would largely cover gas used for heating and cooking, whereas indirect emissions also include those due to household demand for electricity.

Figure 2.2 shows end-user GHG emissions between 1990 and 2011 broken down into six sectors: business (including industrial processes such as cement production), residential, transport (excluding international aviation and shipping), agriculture, public sector and other sectors (waste management, land use and exported energy).

Over the whole period, emissions fell by 29%.[20] However, there was considerable variation across sectors. Business emissions fell by 39%, public sector emissions by 48% and emissions from other sources by 51%. Agriculture emissions fell by 21%. Sectors which are more related to the behaviour of individuals than to that of enterprises saw below-average reductions: residential emissions fell by 23% and transport emissions by 3%.

[20] Note this differs from the change based on Figure 3.1, which includes net sales of ETS emissions.

Figure 2.2: Greenhouse gas end-user emissions by sector, 1990 to 2011

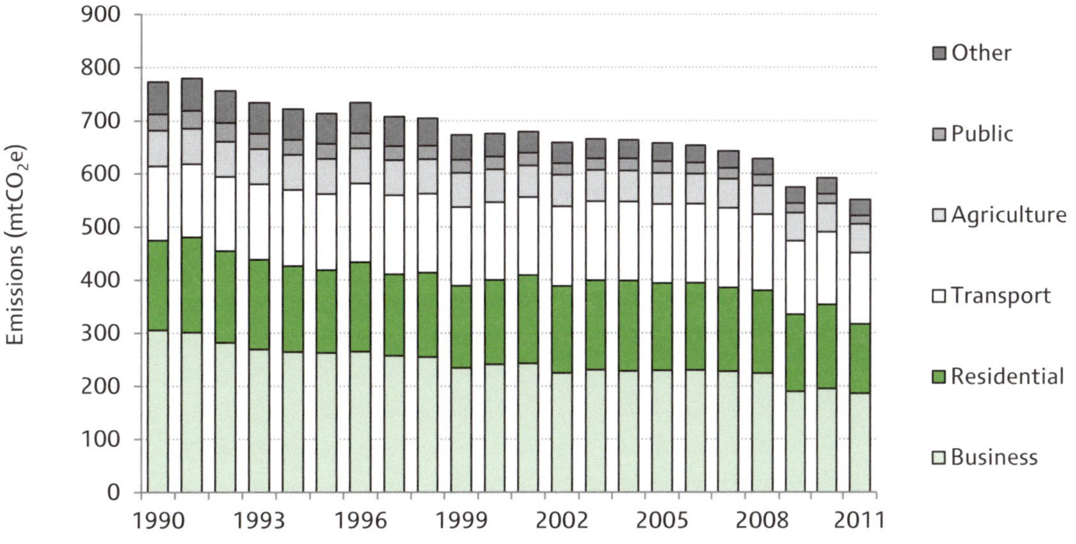

Source: Authors' calculations from DECC 2011 final emissions statistics (https://www.gov.uk/government/uploads/system/uploads/attachment_data/file/73155/2011_final_uk_figures_data_tables.xls, Table 3). Notes: business includes industrial processes. Other includes land use, waste management and energy exports. Transport excludes international aviation and shipping.

There have also been quite strikingly different trends in emissions by source over time. Figure 2.3 shows an index of emissions by the main sectors, set to 100 in 1990. Business and public sector emissions have tended to fall fairly consistently over the whole period (though there was some flatlining in business emissions in the mid-2000s). Agriculture emissions fell between 2000 and 2002, and between 2005 and 2007, but were fairly flat in the remainder of the period. Having fallen during the 'dash for gas' period in the 1990s, residential emissions then *rose* fairly consistently in the early 2000s, and were back at 1990 levels by 2004. Since then residential emissions have fallen more sharply but have been extremely sensitive to weather trends in the past few years. Year-on-year changes in residential emissions were –7% in 2009, +9% in 2010 and –17% in 2011.[21] Transport emissions rose by around 7% between 1990 and 1997 and were then fairly steady for around a decade before falling back fairly markedly since the onset of recession.

[21] The winter months of 2010 were particularly cold, with temperatures in January, February, November and December of 2010 respectively 2.8, 1.8, 1.6 and 5.4 °C below long-term averages since 1971. Source: DECC *Energy Trends* Table 7.1 (https://www.gov.uk/government/uploads/system/uploads/attachment_data/file/69945/et7_1.xls).

Figure 2.3: End-use emissions index by sector (1990 = 100)

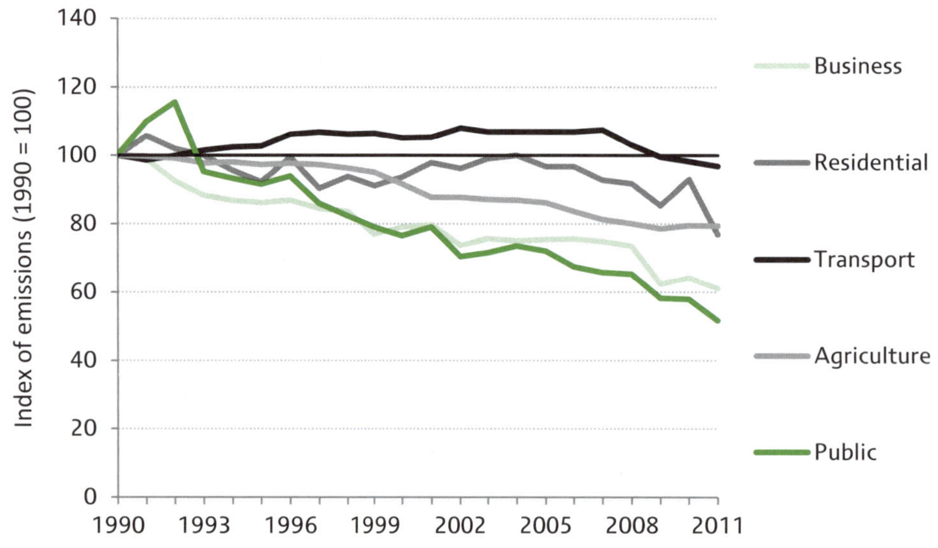

Source: Authors' calculations from DECC 2011 final emissions statistics (https://www.gov.uk/government/uploads/system/uploads/attachment_data/file/73155/2011_final_uk_figures_data_tables.xls, Table 3). Notes: business includes industrial processes. Transport excludes international aviation and shipping.

Figure 2.4: Change in emissions by sector, 1990 to 2011 (%)

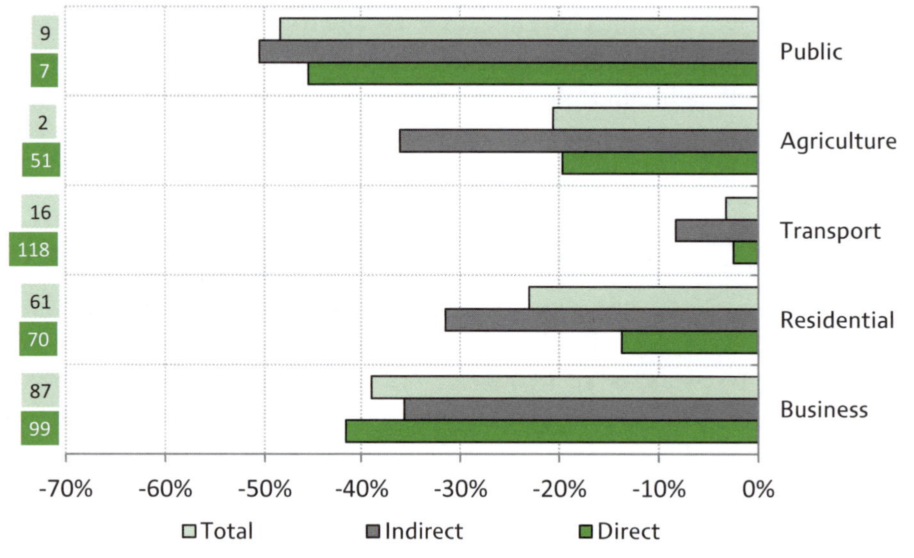

Source: Authors' calculations from DECC 2011 final emissions statistics (https://www.gov.uk/government/uploads/system/uploads/attachment_data/file/73155/2011_final_uk_figures_data_tables.xls, Table 3). Notes: business includes industrial processes. Transport excludes international aviation and shipping. Bars show percentage change between 1990 and 2011; figures show indirect (light box) and direct (dark box) emissions levels by sector in 2011 in $MtCO_2e$.

It is interesting to compare trends not just in end-use emissions but also in direct and indirect emissions within each sector. Emissions from the power sector, for example, fell in the 1990s as generation moved from coal to gas. This would have reduced end-use emissions in other sectors once these emissions were reallocated. Variation in indirect emissions changes across sectors therefore largely reflect changes in electricity demand. Changes in direct (source-based) emissions across sectors also reflect specific behavioural changes within different sectors (reducing the demand for gas, decarbonising production processes and so on). Figure 2.4 therefore breaks down the change in end-user emissions for the five main sectors into changes in direct (source-based) and indirect (re-allocated energy generation) emissions.

Figure 2.5: Change in emissions by sector, 1990 to 2011 (%)

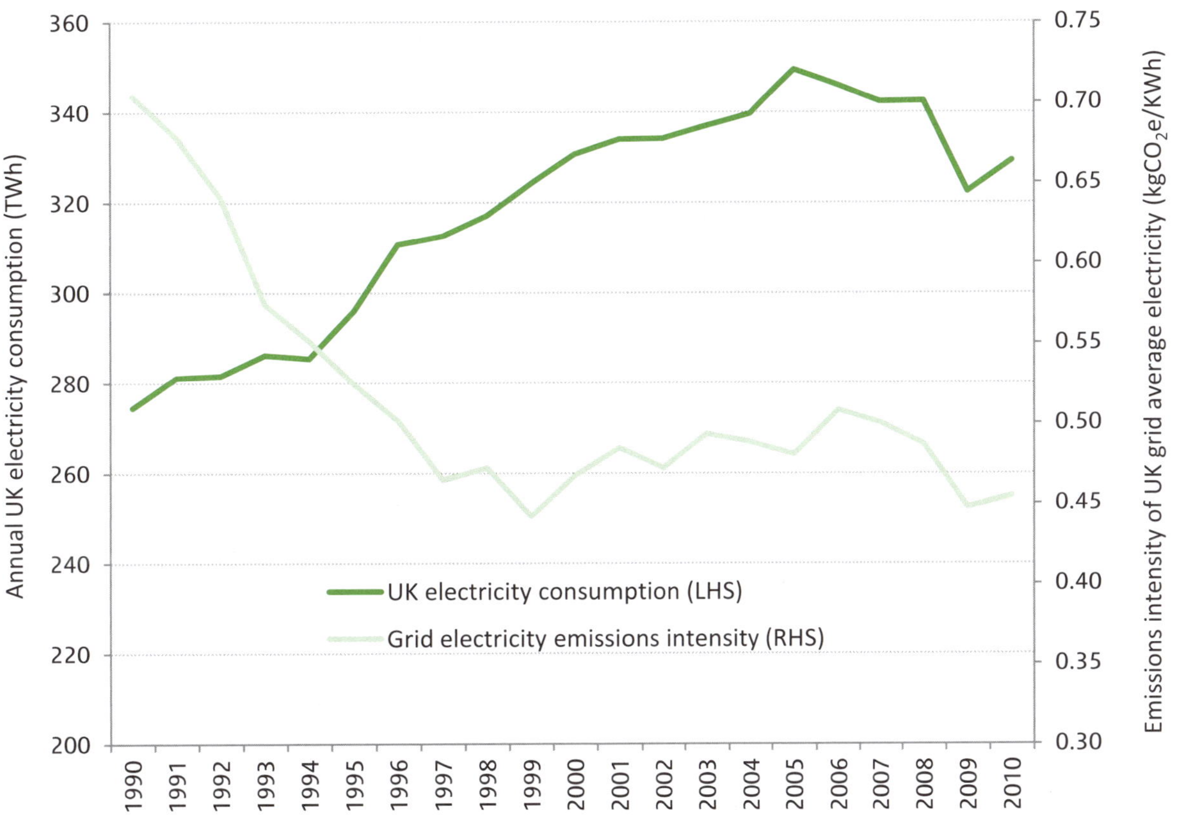

Source: Authors' calculations from DECC July 2013 Digest of United Kingdom energy statistics (DUKES) (https://www.gov.uk/government/publications/electricity-chapter-5-digest-of-united-kingdom-energy-statistics-dukes, table 5.1.2) and DEFRA 2012 GHG emissions factors (https://www.gov.uk/government/uploads/system/uploads/attachment_data/file/69554/pb13773-ghg-conversion-factors-2012.pdf, Table 3a). Notes: UK electricity consumption covers all sectors (domestic, industrial, other) and is given in TWh. The grid electricity emissions intensity shows the kg CO2 per kWh of electricity generated.

Indirect emissions fell by 30% on average with declines seen in all sectors. The largest fall in indirect emissions were seen in the public (–50%), business (–36%) and agriculture (–36%) sectors. The falls in the residential sector (32%) and in particular the transport sector (–8%) were smaller, suggesting that overall electricity demand fell more slowly in those sectors.

It is interesting to note that the fall in indirect emissions occurred even as the amount of energy used in the UK has increased. This reflects the long-term trends in falling emissions associated with electricity generation, first from the dash for gas and, more recently, the increasing share of renewables. Figure 2.5 shows total electricity consumption and the emissions intensity of grid average electricity in the UK between 1990 and 2010. Annual electricity consumption rose from 274.4 TWh in 1990 to 329.3 TWh in 2010, an increase of 20%. Over the same period, the emissions intensity of grid average electricity fell by 35%. This is driven by changes in the generation fuels used. In 1990, 65% of UK electricity was sourced from coal, while only 1% was gas-fired. By 2010, gas accounted for 41% of electricity and coal 32%. At the same time, renewably sourced electricity has increased from a negligible amount in 1990 to over 7% in 2010.

From Figure 2.4, we can also see that variation in the change in direct emissions is even greater than the change in indirect emissions: the largest falls were seen in the public (–46%) and business (-41%) sectors, with smaller falls in agriculture (–20%) and residential (–14%) sectors and a very small fall from transport (–2%). On these disaggregated figures the agricultural sector saw a larger proportional fall in both direct and indirect emissions than the residential sector over this period, but a smaller fall in total emissions. This is because agricultural emissions are more heavily weighted towards direct emissions, which have tended to fall more slowly. In 2011, only 36% of agricultural emissions were indirectly the result of electricity demand compared with 47% of residential.

Broadly, then, the story from Figures 2.2 to 2.4 is that progress in reducing residential and transport emissions has been much slower than in reducing emissions from business, agriculture and the public sector over the past 20 years.

Renewable energy targets

As just discussed, part of the decline in emissions since 1990 has been driven by changes in the fuels used to generate electricity. Meeting future targets is likely to require further reductions in the carbon-intensity of electricity generation. Under the 2009 EU Renewable Energy Directive,[22] the UK has a target to increase the share of renewables in energy use to 15% by 2020. As we detail below, a number of policies have clearly focused on trying to deliver this target. The government

[22] See http://eur-lex.europa.eu/LexUriServ/LexUriServ.do?uri=OJ:L:2009:140:0016:0062:en:PDF.

published its *Renewable Energy Roadmap* in 2011, setting out some of the key technologies that would be critical to meeting the goal.[23]

The overall renewables target does not just cover energy used to generate electricity and for heating, but also energy used in transport (e.g. vehicle fuel). Table 2.2 shows progress towards the target between 2004 and 2011, the most recent year for which figures are available. From a very low base of 1.1% of total energy in 2004, renewables have increased to 3.8% of energy in 2011, but meeting the 2020 target will still require this proportion to almost quadruple within nine years.

Table 2.2: Share of renewables in energy use by sector, 2004 to 2011

Year	Electricity generation	Heating	Transport	Overall
2004	3.5%	0.7%	0.1%	1.1%
2005	4.1%	0.9%	0.2%	1.4%
2006	4.5%	1.0%	0.5%	1.6%
2007	4.8%	1.1%	0.9%	1.8%
2008	5.4%	1.4%	2.1%	2.4%
2009	6.6%	1.7%	2.6%	3.0%
2010	7.4%	1.7%	3.0%	3.2%
2011	8.7%	2.2%	2.9%	3.8%
			2020 target	*15.0%*

Source: Figures to 2006 from Committee on Climate Change (http://hmccc.s3.amazonaws.com/Renewables%20Review/The%20renewable%20energy%20review_Printout.pdf, Table B1). Figures from 2007 onwards from DUKES Table 6.7 (https://www.gov.uk/government/uploads/system/uploads/attachment_data/file/65855/dukes6_7.xls).

Discussion

Given an overall target to reduce emissions, specific targets for renewable energy are in effect subsidiary to that main objective (they also address less well-specified objectives related to energy independence). Setting sub-targets like this risks inefficiency in meeting the overall goal: for example, if it were substantially cheaper at the margin to reduce energy use rather than reduce the carbon intensity of energy generation. Indeed, analysis by DECC suggests that the first three carbon budgets could be met at much lower cost in the absence of the renewables target.[24]

---

[23] See https://www.gov.uk/government/uploads/system/uploads/attachment_data/file/80246/11-02-13_UK_Renewable_Energy_Roadmap_Update_FINAL_DRAFT.pdf.

[24] See, for example, the Analytical Annex to the 2009 Low Carbon Transition Plan, which calculated that the net cost of policies to meet the first three carbon budgets was of the order of £25–29 billion. The cost of support to large-scale renewable electricity alone, however, was calculated at over £31bn, implying that in the absence of this support, carbon budgets could have been achieved at a net negative cost. See http://webarchive.nationalarchives.gov.uk/20100509134746/http://www.decc.gov.uk/Media/viewfile.ashx?FilePath=White Papers/UK Low Carbon Transition Plan WP09/1_20090727143501_e_@@_uklctpanalysis.PDF&filetype=4.

Setting a specific goal for renewable energy could also risk creating a dynamic inefficiency. By anchoring expectations that 15% renewable energy is the 'right' outcome, unexpected shocks to the future costs of renewable energy generation, which might make a different outcome more economically efficient, may not be taken into account unless the policy goal is also adjusted. There is also a risk that when overall and subsidiary objectives are set at different jurisdictions (carbon budgets by the UK government, renewable targets by the EU), they may not be compatible. In the current case, at least in the long term, such incompatibility looks unlikely: achieving a long-term emissions reduction of 80% will not be possible unless most energy is generated free from GHG emissions.

Separate targets can provide additional signals to relevant actors, in particular to renewables investors. This is likely to be important and a degree of certainty can reduce costs to investors. But a balance needs to be struck with the need to avoid spelling out in detail the way in which the target is to be met from the combination of options available. This would have the potential to add considerably to the costs.

Most important is the avoidance of targets that conflict with one another, suggest very different policy objectives and ultimately could lead to a large amount of inefficiency and complexity in policy in order to try to meet them simultaneously. This may well be the case with fuel poverty targets, to which we turn now.

2.2 Reducing fuel poverty

Definition and trends

Fuel poverty is currently defined as a situation in which a household needs to spend at least ten per cent of its income to maintain an adequate standard of warmth in the home.This is defined as 21 °C for the main living room and 18 °C in other rooms. Following the Warm Homes and Conservation Act 2000,[25] the previous government set specific targets to target fuel poverty as part of the 2001 UK Fuel Poverty Strategy.[26] The first objective was to eliminate fuel poverty in all 'vulnerable' households by 2010 'as far as is reasonably practicable'.[27] The second objective was to eliminate fuel poverty in all households by 2016.

Figure 2.6 shows that initially clear progress was made towards the targets. Fuel poverty rates (shown as the bars and read against the left-hand axis), which had declined sharply prior to the introduction of the targets in 2001, continued to fall, reaching a trough of around 1.2 million households in 2003 and 2004. The trend then reversed markedly: fuel poverty levels among the whole population more than trebled by 2009, before falling back in 2010 and 2011. The 2010 target for

[25] http://www.legislation.gov.uk/ukpga/2000/31/contents.

[26] http://webarchive.nationalarchives.gov.uk/+/http://www.berr.gov.uk/files/file16495.pdf.

[27] 'Vulnerable' households are those which include members who are disabled or have a long-term illness, elderly householders (over the age of 60) or households containing children.

vulnerable households was missed, with 2.8 million vulnerable households classified as fuel poor in England in 2010.

Figure 2.6: Households in fuel poverty in England, 1996 to 2011

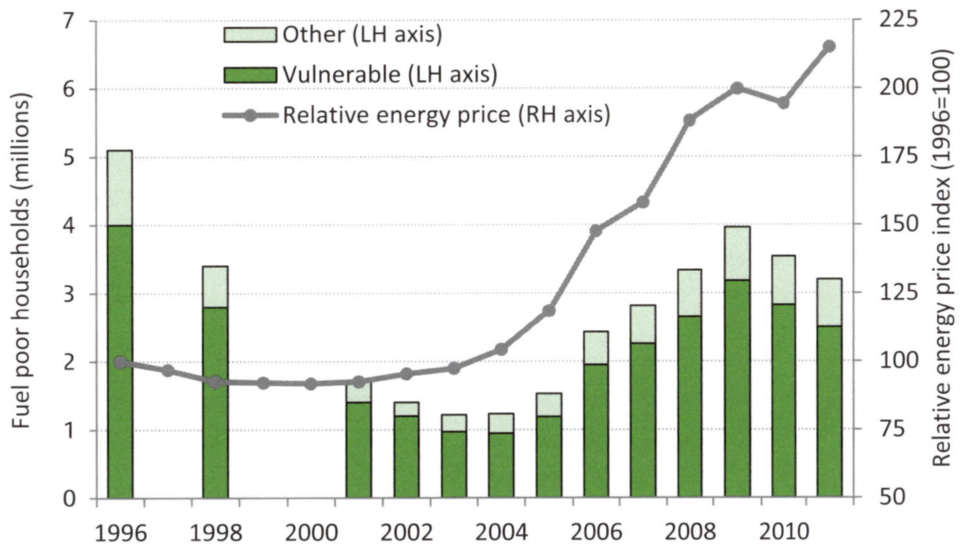

Source: Figures before 2003 from DECC Fuel Poverty Monitoring Indicators 2012 (https://www.gov.uk/government/uploads/system/uploads/attachment_data/file/66017/5272-fuel-poverty-monitoring-indicators-2012.pdf). Figures from 2003 onwards from DECC Trends in Fuel Poverty in England 2003 to 2011 (https://www.gov.uk/government/publications/trends-in-fuel-poverty-england-2003-to-2011). RPI index is authors' calculations from ONS data. Note: estimates of fuel poverty rates not available in all years.

As shown in Figure 2.6, a key driver of fuel poverty is the price of energy. The line, read against the right-hand axis, shows the relative price of household energy compared with all prices from the Retail Prices Index each year, indexed at 100 in 1996. Continuing a trend which began in the mid-1990s, relative energy prices fell by almost 10% between 1996 and 2000, but then rose rapidly, more than doubling by 2011. Trends in fuel poverty reflect this price variation, though it is notable that despite the very large price increases, fuel poverty levels were lower in 2009 at their recent peak than they were in 1996, and that fuel poverty levels fell in 2011 despite another up-tick in relative energy prices. This may be explained, at least in part, by improvements in the energy efficiency of the homes of low income households across the period. Hills (2012) shows that the number of the poorest 30% of households living in the least energy efficient properties (SAP bands E, F and G) fell from 4.5 to 3 million households (a decrease of 33%) between 1996 and 2009, and suggests that a steady fall in fuel poverty would have been expected to occur in the absence of these large price rises, Recognising that the sensitivity of fuel poverty levels on the '10%' definition to energy prices made it virtually impossible to hit targets which had been set in an era of much lower prices, the government commissioned a review of the definition (see Box 2.1). Department of Energy and Climate Change (2013a) have indicated they will adopt the new definition proposed by the review, known as the 'low income, high costs' measure, along with a new fuel poverty target (still to be formalised) based around improving the energy efficiency of homes for those found to be fuel poor under the new measure.

Box 2.1: The Hills Review of the definition of fuel poverty

In 2011, DECC commissioned a review of fuel poverty to be led by Professor John Hills. The terms of reference of the review were to explore whether fuel poverty was a distinct issue for policy concern, how fuel poverty should be measured and what the policy implications of changing the measure of fuel poverty might be.

The interim review (Hills, 2011) concluded that fuel poverty was a distinct concern and the final review (Hills, 2012) argued for a change in the way that fuel poverty was measured. Rather than defining a household as fuel-poor if their required energy costs exceeded 10% of income, Hills proposed two criteria:

1. Households have required fuel costs that are above average (defined as the median across the population adjusted for household composition);
2. Households are below the income poverty threshold (defined as 60% of median income adjusted for household composition), based on a measure of income which removes required fuel costs as well as other housing costs.

Hills also recommended an indicator of the *depth* of fuel poverty, based on the total amount by which the required fuel costs of all fuel-poor households exceeded the population median. This he defined as the 'aggregate fuel poverty gap'.

The Figure below shows DECC estimates of fuel poverty in England on both of the measures from the Hills definition, and compares them against the 10% definition as shown above. There are strikingly different trends in the two Hills measures: the number of fuel-poor households barely fell in the mid-2000s and rose more slowly in the late 2000s. The *depth* of fuel poverty rose fairly consistently from the mid-2000s onwards, however. Both Hills measures are far less volatile than the previous 10% measure. This difference highlights how strongly the current definition is influenced by energy prices. In the Hills definition, by contrast, higher prices have a stronger impact on the *depth* of fuel poverty but not necessarily on its overall incidence.

Figure: Fuel poverty measured by the Hills criteria, England, 1996 to 2011

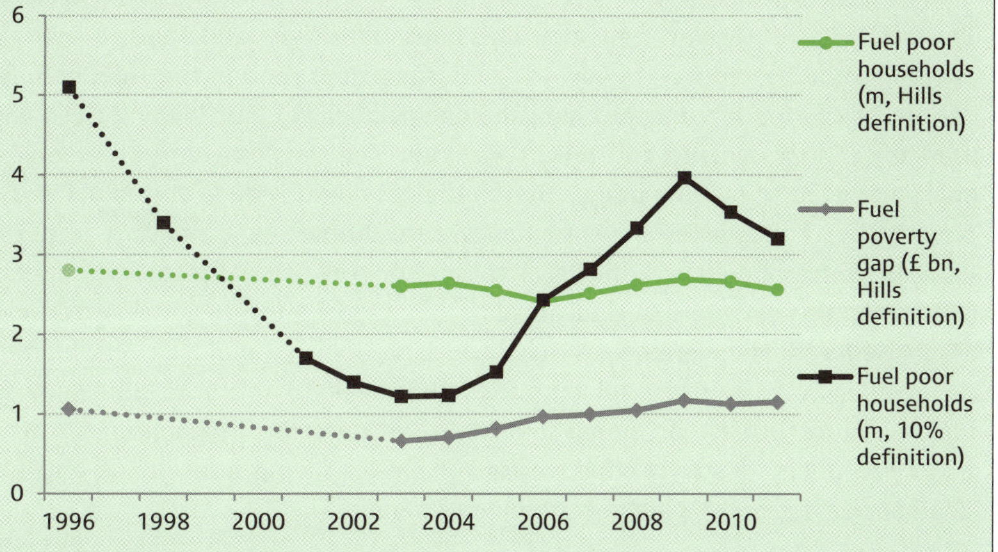

Source: Department of Energy and Climate Change (https://www.gov.uk/government/organisations/department-of-energy-climate-change/series/fuel-poverty-statistics).

To the extent that we believe it is important to measure 'fuel poverty' at all (see Chapter 2.2), in general there are reasons to prefer the Hills measure to the 10% measure. Choosing 10% as the threshold was always somewhat arbitrary. Large fluctuations in the headcount fuel poverty levels in the face of higher energy prices suggest many households were around that level and pushed above or below the line as prices changed, even though for all practical purposes it is hard to see why someone needing to spend 9.9% of their money on fuel is not 'fuel poor' whilst someone needing to spend 10.1% is. The headcount Hills measure is much less sensitive to price since it depends on average within-year energy spending, and will not count relatively well-off people as 'fuel poor' if they happen to face (presumably as a matter of choice given their means) very high energy costs as a result of their housing conditions. The headcount Hills measure also does not favour reducing energy bills as a policy response over reductions in overall income poverty, whereas the policy incentive under the 10% measure was always to reduce energy bills, particularly for those who were just over the threshold, as a way to maximise the impact on measured fuel poverty. Combining a headcount and depth measure also acknowledges this latter point, since policies which reduce the depth but not necessarily overall incidence of fuel poverty will at least 'show up' in the measured statistics.

Discussion

There is a clear tension between emissions objectives and fuel poverty. Fuel poverty could be reduced by lowering energy prices (and in the Hills definition, energy prices for low income households specifically), but lower prices would increase energy demand and raise emissions. The tension could be resolved by energy efficiency: making homes more efficient could reduce emissions and fuel poverty by reducing the costs of heating the home. As a result it is perhaps unsurprising that policies which seek to raise energy prices (ensuring that energy users face the correct marginal incentives including the negative externalities associated with their energy use choices) have focused almost overwhelmingly on business energy use whilst policies that seek to promote efficiency have focused more heavily on households.[28] As we describe in Chapter 6, this has resulted in very different effective carbon prices facing households and firms. The policies themselves are described in Appendix A.

There are of course reasons to question the rationale for a fuel poverty target at all. The government is concerned about equity overall, but why in energy policy specifically?

Hills (2011) argued that fuel poverty is a distinct phenomenon for three reasons:

1. Low income households face higher costs of heating their home, in part because they tend to live in more inefficient houses and in part because they use more expensive energy payment methods. There is relatively little that poor people can do to mitigate these problems (for example, they may not be able to afford to pay for efficiency improvements or to replace prepay energy meters).
2. Being unable to heat the home adequately is associated with adverse health outcomes.
3. Fuel poverty acts as a barrier to implementing policies to reduce carbon emissions because price increases will have adverse distributional effects.

These are important points and almost certainly provide some rationale for helping poorer households with energy efficiency measures for example. But it remains questionable whether they imply a separate fuel poverty target.

The first point is not dissimilar to the argument that housing costs should not be included in the measure of income used to measure poverty. We have an official poverty measure 'after housing costs'. Perhaps an additional 'after fuel costs' measure would be helpful.

[28] It is worth noting that there has been recent interest in the idea that energy efficiency improvements lead to 'rebound effects' (see, for example, Chitnis et al, 2013) whereby some of the expected reduction in energy demand following an improvement in efficiency is offset. This can occur because the increase in efficiency effectively reduces relative energy prices, or because some of the income saved is spent on energy through some 'comfort-taking' effect, or because there is some signalling or salience effect on energy demand from installing efficiency measures. Possible rebound effects are another reason to want to ensure that energy users face price incentives which properly account for the various external costs associated with energy use: making homes more efficient without also ensuring that households face the full carbon costs of their energy choices might simply exacerbate the rebound effect.

The second point can be made in other contexts as well. For example, there are costs to health associated with poor nutrition, and poorer households may have less scope to meet nutritional standards if they are costly to attain. This does not imply we also need to monitor 'food poverty'.[29]

The final point is something of a circular argument: fuel poverty is a barrier to implementing policies to reduce emissions only because fuel poverty is seen to be a distinct problem from income poverty. There may well be adverse distributional consequences from policies which raise energy bills and prices, but as we argue in Chapter 8 and in our companion report (Advani et al., 2013), these could be mitigated through other changes to the tax and benefits system.

2.3 Increasing the share of revenues from green taxes

The coalition agreement signed in 2010 set out an explicit objective to 'increase the proportion of tax revenue accounted for by environmental taxes'.[30] This objective is therefore relevant in thinking about energy-use policies, assuming that any taxes on energy use would be counted as 'environmental'.

It was not until July 2012, more than two years after the coalition was formed, that a formal definition of the pledge was announced by the Treasury.[31]

1. 'Environmental' taxes are interpreted only as those whose primary objective is to encourage pro-environmental behaviour. On the government's interpretation this encompasses just five measures. Two (aggregates levy and landfill tax) are primarily focused on the use of natural resources, whilst three (the carbon reduction commitment, climate change levy and EU emissions trading auction revenues) are primarily energy-related taxes.
2. These taxes should make up at least the same proportion of revenue in 2015–6 as they did in 2010–11.

Other definitions of 'environmental taxes' encompass taxes which result in pro-environmental *outcomes* even if that is not the only explicit *intent* of the tax. For example, the ONS Environmental Accounts include taxes on vehicle fuel, air passengers and vehicle purchases as 'environmental'.[32]

Leicester and Stoye (2012) assessed whether the coalition pledge would be met on the government's own definition of the set of environmental taxes, the ONS

---

[29] Although such measures are being considered: see, for example, http://www.food.gov.uk/northern-ireland/nutritionni/ninutritionhomeless#.UajYD1ewVDA.

[30] See https://www.gov.uk/government/uploads/system/uploads/attachment_data/file/78977/coalition_programme_for_government.pdf.

[31] See https://www.gov.uk/government/news/definition-of-environmental-tax-published.

[32] See http://www.ons.gov.uk/ons/guide-method/method-quality/specific/economy/environmental-accounts/monetary-accounts/government-revenues-from-environmental-taxes.pdf.

definition and a third definition which broadly encompassed the other two. They found that on the government's definition, the pledge was met with ease: the green tax share is estimated to more than double, from 0.4% to 0.9% of receipts by 2015–6. On any wider definition, however, the pledge is missed. Using the Leicester and Stoye (2012) classification, for example, green taxes fall from 7.3% of revenues in 2010–11 to 7.0% in 2015–6.

The key difference is that new or increased energy taxes, which account for most of the official definition, are more than offset in wider definitions by a fall in the share of revenue coming from vehicle fuel duties. Based on the most recent Office for Budget Responsibility forecasts alongside the March 2013 Budget, the three energy taxes in the official measure are set to rise from 0.1% of receipts in 2010–11 to 0.6% in 2015–6. However, fuel duties are estimated to fall from 5.0% to 4.1% over the period. This follows a number of concessions by the Chancellor since coming to office, in which nominal duties were frozen or reduced at successive fiscal events. If fuel duties are not classified as 'environmental taxes' then, oddly, reducing them actually makes the pledge *easier* to meet since it reduces 'non-environmental' revenues.

Making pledges which are subsequently defined in a way which makes them trivial to meet (given inherited plans for increased energy taxes) and in terms of their importance (given the very minor role for green taxes on the official measure) does not lend them much credibility.

3. Revenues and expenditures on current UK energy use policies

Arun Advani, Samuela Bassi, Paul Johnson, Andrew Leicester and George Stoye

This chapter discusses the current set of policies which affect energy use in the UK. To summarise the current landscape, we begin with a brief overview of the measures, grouped according to their main overall objective (more policy details are provided in Appendix A). We then set out estimates of how much revenue is raised from the policies that act as effective taxes on energy use, and compare them with estimates of the amount spent on other policies that effectively subsidise energy use for particular groups or act as revenue transfers in other ways. We then offer some comments on current policies in three broad groups: those which affect upstream energy prices, those which affect downstream energy used by firms and industry and those which affect downstream energy used by the household sector.

3.1 Current policies related to energy use

Table 3.1 provides a brief summary of policies currently in place, or those that will be implemented in the near future, which are related to energy use. More detail is provided on the design of the policies and evidence of their impact or effectiveness in Appendix A.

Table 3.1: Summary of current policies related to energy use

Policy	Description
Policies focused on reducing emissions through pricing carbon	
EU Emissions Trading Scheme (ETS)	A cap-and-trade scheme for direct CO_2 emissions from energy-intensive facilities, introduced in 2005. A 'cap' specifies the total number of emissions allowed, set at 2,039 $MtCO_2$ across the EU in 2013 for 'fixed' installations (such as power plants and other industrial installations). This will be reduced annually by around 37 $MtCO_2$. In 2020 allowed emissions will be around 1,777 million, 21% lower than in 2005. Permits are allocated to participants via a mixture of free allocation and auctions, and are traded to establish a carbon price. Revenues from auctioned permits in the UK are forecast to be £700 million in 2013–4.
Carbon Price Floor (CPF)	A tax (the Carbon Price Support Rate, CPSR) on fuels used for electricity generation, set so the combined carbon price including the ETS meets an increasing trajectory (£16/tCO_2e in 2013 rising to £70 in 2030, in 2009 prices). The CPSR is set two years in advance, and

	varies by fuel according to carbon content. The CPF is expected to reduce emissions by 261 $MtCO_2$ by 2030. CPSR revenues of £740 million are forecast in 2013–4.
Climate Change Levy (CCL)	A tax, introduced in 2001, levied on the supply of electricity, gas, liquified petroleum gas (LPG) and solid fuels supplied to businesses. Rates vary across energy types but do not reflect differences in fuel carbon content. 2007 estimates projected annual savings of 3.5 mtC or 12.8 $MtCO_2$ in 2010 as a result of the CCL (National Audit Office, 2007). CCL receipts were £636 million in 2012–3.
Climate Change Agreement (CCA)	CCAs provide a discount on the rates of CCL (currently 90% for electricity and 65% for other fuels) to certain energy-intensive industries in exchange for agreements to undertake actions to reduce carbon emissions. These discounts are estimated to have cost £170 million in 2012–3.
Carbon Reduction Commitment Energy Efficiency Scheme (CRC)	Relatively large firms and public sector organisations that are not direct participants in the EU ETS are required to report on their electricity and gas consumption. This is converted to implied carbon values. Firms must have purchased sufficient allowances (currently priced at £12/tCO_2) to cover their emissions. Revenues are estimated at £700 million in 2013–4.
Emissions Performance Standard (EPS)	Mandatory emissions standards for new fossil fuel power stations, imposing an annual limit on the total CO_2 emissions per unit of installed capacity that they are allowed to emit. The standard is 450 gCO_2/kWh from 2014.
Policies focused on reducing emissions through support for low-carbon energy	
Renewables Obligation	Launched in 2002, the RO requires energy suppliers to source an increasing proportion of electricity from renewable sources. Purchasing renewable energy gives suppliers Renewable Obligation Certificates (ROCs), which can be traded. Suppliers who hold insufficient ROCs must buy out their remaining requirement. The combined value of the ROCs presented for compliance in 2011–2 was £1.45 billion.
Contracts for Difference with Feed-in Tariffs (CfD FITS)	Generators of renewable energy agree long-term contracts to supply energy at a 'strike price'. If the wholesale market price is below this, the generator receives the difference from the contract counterparty; if the market price is higher the generator pays the difference to the counterparty. The policy will be launched in 2014, and will replace the RO for new generation by 2017.

Small-scale Feed-in Tariffs (FITs)	Payments made to households and businesses who install small-scale renewable generation technologies (up to 5 MWh). First introduced in 2010, the payments vary by technology and date of installation. Additional payments may be made for energy exported to the National Grid. Payments are guaranteed for a minimum of 20 years. The FITs budget is £328 million for 2013–4.
Renewable Heat Incentive (RHI)	Payments made to firms (and households from 2014) who installed renewable heat technologies. Payments vary by technology and scale, and are paid on a quarterly basis over a 20-year period. The scheme was introduced in April 2011, and has a budget of £251 million in 2013–4.
Capacity Mechanism	The Capacity Mechanism aims to ensure sufficient capacity to meet future demand. Four-year forecasts of peak energy demand are made to calculate required capacity. Capacity is contracted through a competitive auction, where the winners commit to provide electricity when required. In return, generators are paid for available capacity (as opposed to the actual energy supplied). Generators face penalties if unable to provide the promised capacity The first auction is set to take place in 2014, with an initial delivery year in 2018–9.
Policies focused on improving energy efficiency	
Carbon Emissions Reduction Target (CERT)	CERT was an obligation on large energy suppliers to improve the efficiency of the housing stock through insulation and other efficiency measures, in place from 2008 to 2012. At least 40% of savings needed to be made among a priority group (including households with elderly individuals or households in receipt of a number of means-tested benefits). CERT led to the installation between 2008 and 2012 of measures estimated to achieve lifetime carbon savings of 293 $MtCO_2$.
Community Energy Saving Programme (CESP)	CESP was an obligation on large energy suppliers and generators to reduce residential emissions in areas of high deprivation. Qualifying measures included solid wall insulation and replacing inefficient boilers; savings from cavity wall and loft insulation were tightly restricted. Suppliers achieved bonuses for providing multiple measures to the same household or assisting a large fraction of households in an area. CESP achieved an estimated lifetime emissions reduction of 19.25 $MtCO_2$ through the installation of measures between 2009 and 2012.
Energy Company Obligation (ECO)	ECO replaced CERT and CESP from 2013. It requires large energy companies to support domestic energy efficiency through three main components. The 'home heating cost reduction' focuses on support (worth £4.2

	billion) for poor and vulnerable households; the 'carbon saving community obligation' requires savings equivalent to 6.8 $MtCO_2$ in deprived areas; and the 'carbon emissions reduction obligation' requires savings of 20.9 $MtCO_2$ from hard-to-treat properties and expensive measures that would not be covered by the Green Deal (see below).
Green Deal	The Green Deal provides loans to finance efficiency and insulation measures. Following an assessment and agreed installation of measures, the loan is repaid through the energy bill, tied to a property rather than an individual. Only measures deemed to meet the 'Golden Rule' (that the amount repaid in the first year must be no more than the expected bill saving from the measure) can be installed under the Green Deal. Loans with a forecast value of £3.2–4.1 billion are expected to be in place by 2022.
Products Policy	The Products Policy encompasses a range of measures that affect the efficiency of energy-using products and appliances. This includes the EU 'Ecodesign' Directives which impose increasingly stringent minimum energy standards on different product groups. The policy also covers energy-related labelling and building regulations, and is expected to save 14.3 $MtCO_2$ per year by 2020.
Smart Meter Rollout	Smart meters record information on energy use which is transmitted directly to energy suppliers without the need for physical visits to read meters. Real-time information on energy usage will be provided to consumers through in-home display units. This should aid energy users to reduce energy demand bills. The Smart Meter Rollout is currently in its 'Foundation Stage', with common technical standards and a regulatory standard being agreed. The roll-phase will begin in 2015.
Policies giving support to energy bills	
Warm Home Discount (WHD)	An obligation on large energy suppliers to provide electricity bill rebates, worth £135 in 2013–4, to low income and vulnerable households. Those on the Guarantee credit element of pension credit (or just the savings credit element for older consumers) receive automatic rebates. Energy companies can set their own rules about which other vulnerable groups can apply for a rebate, typically those on means-tested benefits with young children or a disabled member. WHD was introduced in 2011–2, and has a budget of £320 million in 2013–4, part of which covers other support and advice to reduce bills and some legacy costs of providing cheaper tariffs to vulnerable households.

Winter Fuel Payment (WFP)	A cash transfer, initially introduced in 1997, to households containing someone over the female state pension age. In 2013–4 the payment is £200, rising to £300 if someone is aged 80 or over. Payments are at the household level and split if there are multiple eligible individuals. In 2012–3, 12.7 million payments were made at a cost of £2.15 billion.
Cold Weather Payment (CWP)	A cash transfer, introduced in 1986, to vulnerable households to meet the cost of higher energy bills in periods of cold local weather. The payment is currently £25 following every seven-day period in which temperatures at a nearby weather station are forecast to fall below 0 °C. Eligible households include those in receipt of a range of means-tested benefits with older people, young children or disabled people. In 2012–3, 5.8 million payments were made at a cost of £146.1 million.
VAT on energy use	Residential customers pay a VAT rate of 5% on domestic energy use (including electricity, gas and non-metered fuels such as coal) compared with a standard VAT rate of 20%. Energy used by non-domestic customers is taxed at the standard rate of 20%. The cost of the reduced VAT rate for domestic energy is estimated at around £5.2 billion per year.

Revenues and expenditures relating to energy use

How much revenue is raised in the UK from taxes related to energy use? Table 3.2 shows forecasts of receipts from energy-related taxes between 2012–3 and 2017–8 based on Office for Budget Responsibility (OBR) figures consistent with the March 2013 Budget. We include estimates for the Renewables Obligation, small-scale Feed-in Tariffs and the Warm Homes Discount. These policies are administered by energy companies with the costs recouped through energy prices and bills. The revenues are not raised and spent by government directly, but are treated as part of tax revenue and government expenditures by the OBR in its forecasts. This follows the development of the 'levy control framework' between DECC and HM Treasury established in the 2010 Spending Review (see Box 3.1).

Table 3.2: Actual and forecast revenues from energy-related taxes, 2012–3 to 2017–8

£ billion	2012–3	2013–4	2014–5	2015–6	2016–7	2017–8
Climate Change Levy	0.7	1.5	2.0	2.5	2.5	2.5
EU ETS auctions	0.3	0.7	0.7	0.8	0.8	0.9
Carbon Reduction Commitment	0.7	0.7	0.9	0.9	1.0	1.0
Renewables Obligation	0.5	0.6	0.7	0.9	1.1	1.4
Small-scale FITs	0.5	0.6	0.8	1.0	1.1	1.3
Warm Homes Discount	0.3	0.3	0.3	0.3	0.3	0.3
Total energy-related	3.0	4.4	5.4	6.4	6.8	7.4
Total taxes	586.8	612.4	633.1	657.6	694.1	723.0
Energy-related %	*0.5%*	*0.7%*	*0.9%*	*1.0%*	*1.0%*	*1.0%*

Source: Office for Budget Responsibility Economic and Fiscal Outlook, March 2013 (http://budgetresponsibility.independent.gov.uk/pubs/Copy-of-March-2013-EFO-charts-and-tables.xls, Table 4.7 and http://budgetresponsibility.independent.gov.uk/pubs/Economic-and-fiscal-outlook-supplementary-fiscal-tables-March-2013.xls, Table 2.7). Note: Climate Change Levy includes the Carbon Price Support Rate, and is net of any discounts available under Climate Change Agreements.

In total, energy-related taxes are forecast to raise around £4.4 billion in the current financial year, or just over 0.7% of receipts. This has increased from £3.0 billion (0.5%) in 2012–3. This will rise to £7.4 billion in cash terms by 2017–8, a fraction over 1% of total revenue.

These policies essentially increase the cost of energy for households and firms. For example, Department of Energy and Climate Change (2013b) estimates that the cost of the Renewables Obligation, EU ETS, small-scale FITs and the Warm Home Discount accounted for £56 (4.4%) of the average household dual-fuel energy bill in 2013.[33] However, the government also spends substantial sums of money on policies which reduce energy costs for some households and firms. These effective subsidies can take the form of reduced tax rates (so-called 'tax expenditures') or direct support for energy bills.[34] Their total value is much larger than the revenues raised from the taxes listed in Table 4.2. The measures include:

[33] These figures reflect only the costs of these policies, and do not reflect the overall impact on bills. DECC (2013X) estimate that, through reductions in energy consumption, climate change and energy policies reduced the average household dual-fuel energy bill by £15 (1%) relative to a scenario where no climate change policies were in place.

[34] Note that we do not include here policies or tax expenditures which subsidise or support renewable energy; the focus is on policies which effectively subsidise the costs of non-renewable energy through either reduced

- **Reduced-rate VAT on domestic energy** (electricity, gas and other fuels) at a cost of £5.2 billion in 2012–3.
- **Reduced Climate Change Levy for energy-intensive firms** at a cost of £170 million in 2012–3.[35]
- **Winter Fuel Payments** which appear to be spent far more on energy than an unlabelled cash benefit (Beatty et al., 2011) and so could be thought of as *de facto* energy bill subsidies. These cost £2.2 billion in 2012–3.[36]
- **Cold Weather Payments** which again are not necessarily spent on energy, but given both the label and the conditional payment in periods of very cold weather, are likely to be heavily spent on heating. These cost £146 million in 2012–3.[37]

Box 3.1: DECC levy-control framework

In the 2010 Spending Review, HM Treasury and DECC agreed a 'control framework' that set a cap on the overall value of policies which support DECC objectives on climate change and fuel poverty, which are paid for by energy companies, and where the costs are recouped through consumer energy bills. There is considerable uncertainty about *how* energy companies actually recoup these costs: as higher per-unit energy prices or through fixed levies on bills (e.g. higher standing charges). The implications of the two are quite different. Though both would be regressive (having a greater relative impact on poorer households who devote more of their budget to energy costs), a fixed levy would be much more regressive (akin to a poll tax) than a per-unit charge, since richer households consume more energy (see Advani et al., 2013).

Policies in the framework can be seen as implicit tax and spend measures: rather than spending tax revenues on measures to support policy objectives, the spending is done by private energy companies who are allowed to recoup the costs through unavoidable levies on energy bills and energy prices. Since tax and spend measures fall under the responsibility of HM Treasury, the intention was to set out limits for how much money could be spent on these measures (whilst not affecting any payments which had already been guaranteed) in the same way that limits on

prices or direct energy bill support. We also exclude policies which are delivered through energy suppliers such as Warm Home Discount and ECO, which reduce bills for some people but are recouped through higher energy bills for others. The focus is on policies which are paid for through general taxation.

[35] Excluding the costs to affected firms of meeting their Climate Change Agreements.

[36] See http://statistics.dwp.gov.uk/asd/asd4/budget_2013_260313.xls. Beatty et al. (2011) argue that around 41% of the WFP is devoted to energy compared to 3% from an unlabelled transfer. Even if we only treat 40% or so of WFP as effectively a bill subsidy, that still amounts to almost £900 million.

[37] See https://www.gov.uk/government/uploads/system/uploads/attachment_data/file/198774/sf_cold_payts_eoy_201213.pdf.

spending financed through general taxation are set out for departments in Spending Reviews.

The costs of measures in the control framework are likely to be included as taxes and expenditures in public finance measures, pending ONS decisions.[a] This would mean funding measures through the framework has the same public finance implications as funding them through general taxation.

Three policies are currently part of the control framework:

- Renewables Obligation (eventually to be replaced by CfD FITs)
- Small-scale FITs
- Warm Homes Discount

The agreed amount to be spent on each policy within the framework over the period to 2014–5 is:

£ million	2011/12	2012/13	2013/14	2014/15
RO	1,750	2,156	2,556	3,114
FITs	94	196	328	446
Total renewable	***1,844***	***2,352***	***2,884***	***3,560***
WHD	250	275	300	310
Grand total	**2,094**	**2,627**	**3,184**	**3,870**

Source: Department of Energy and Climate Change (https://www.gov.uk/government/uploads/system/uploads/attachment_data/file/48244/3290-control-fwork-decc-levyfunded-spending.pdf).

DECC are required to produce forecasts of the cost of policies within the framework, agreed with HM Treasury and the OBR, which ensure the total cost does not exceed the agreed cap. If the forecast cost (or out-turn spend) is above the cap then DECC have to provide a plan to reduce spend to the capped value, though if the overshoot is relatively small (HM Treasury agreed an initial 20% headroom above the cap to account for unanticipated shocks) and unlikely to persist then it may be that no action is required. If DECC are unable to formulate a plan to reduce spending to the cap, then the excess may be required to be taken from DECC's general departmental spending. Any underspend in the cap cannot be carried forward to future years.

It is the total figure rather than the policy-by-policy figure which is set under the control framework. New policies which were similarly levy-funded would have to be offset by reductions in spending on existing policies unless there was an agreement to increase the cap. For example, further support for efficiency measures for electricity use, currently being consulted on by DECC,[b] might be paid for in this way. There is also the possibility that Energy Company Obligation (ECO), not currently part of the control framework but similarly paid for through higher customer bills, will also be classified as imputed tax and spend by the ONS.[c] The cost of ECO for energy companies could be around £1.3 billion per year, around one-third of the total cap for 2014–5.[d]

In real terms (2011–12 prices) the overall cap is set to rise from £3.3 billion in 2014–5 to £7.6 billion by 2020-21.[e] The increase in the cap to 2020/21 suggests that support for renewables after 2014–5 will increase at roughly the same rate as before.

[a] Note that the ONS have so far only agreed that the RO should be classified as a tax and spend measure. Decisions are still pending on the treatment of WHD and FITs. Should these policies not be classified as tax and spend measures (but instead be treated as e.g. regulation) it is not clear how this would impact the control framework. However it is seen as likely that they will be treated as such and forecasts for them are included in OBR figures for future revenues under this assumption.

[b] See https://www.gov.uk/government/uploads/system/uploads/attachment_data/file/66561/7075-electricity-demand-reduction-consultation-on-optio.pdf.

[c] See https://www.gov.uk/government/uploads/system/uploads/attachment_data/file/42984/5533-final-stage-impact-assessment-for-the-green-deal-a.pdf, footnote 76.

[d] See https://www.gov.uk/government/uploads/system/uploads/attachment_data/file/42984/5533-final-stage-impact-assessment-for-the-green-deal-a.pdf, table 14.

[e] See https://www.gov.uk/government/uploads/system/uploads/attachment_data/file/209361/Levy_Control_Framework_and_Draft_CfD_Strike_Prices.pdf.

Even if we exclude the labelled payments (WFP and CWP), the total subsidy in 2012–3 was almost £5.4 billion, compared with revenues from energy related taxes of £3.0 billion. Including them, the subsidy rises to almost £7.7 billion.

Tax revenues are not the only transfers created by energy policies. Several firms benefit from the free allocation (grandfathering) of allowances under the EU ETS. In the UK, about 230 million allowances were allocated for free in 2012, with a total asset value of almost £1.4 billion (European Environment Agency , 2013), applying the average spot price across the period (€6/tonne CO_2). The value was considerably higher in phase I of the scheme. In phase III of the EU ETS, running from 2013 to 2020, the number of auctioned allowances has increased to 50%, compared with 10% in Phase II (2008–12). This reduces the size of rents, but a significant amount of free allocations remain in place. For instance, had only half the EU allowances (EUAs) been grandfathered in 2012, the asset value of the free allowances would still have been of the order of £0.7 billion.

Generators of renewably sourced electricity also receive substantial transfers from energy suppliers. These transfers occur through two main policies: the Renewables Obligation and Feed-in Tariffs. These expenditures are capped under the Levy-control framework (see Box 3.1) and are ultimately recouped through higher electricity bills. In 2011–12, these policies accounted for transfers of £1.5 billion to large-scale renewable generators through the Renewables Obligation

(Ofgem, 2013a) and £140 million to small-scale generators through Feed-in Tariffs (Ofgem, 2013b). These transfers are set to increase to £2.8 billion (Department of Energy and Climate Change, 2012a) and £860 million (Department of Energy and Climate Change, 2012b) respectively by 2014–5 (2011–12 prices).[38]

Policies aimed at reducing emissions through changing the price of electricity create transfers to non-marginal producers of electricity who do not bear the costs of the policy. For example, electricity prices are set by marginal power plants, which are assumed to be Combined Cycle Gas Turbines (CCGT). Prices therefore reflect the cost of the ETS and the CPF on these generators. The same price is received by non-marginal generators, despite the fact that these generators do not bear the cost of these policies, such as nuclear or renewable plants. The ETS and CPF added approximately £3/MWh to average retail prices in 2013 (Department of Energy and Climate Change, 2013b). This equates to transfers of £192 million to nuclear plants and £127 million to renewable generators in the same year.

[38] Forecast transfers arising from the FIT scheme are given in nominal terms in DECC (2012X). These have been converted to 2011/12 prices using HMT June 2013 GDP deflator.

4. Current UK upstream and firm energy use policies

Samuela Bassi, Alex Bowen and Sam Fankhauser

4.1 Upstream carbon prices

Currently the UK electricity generation sector is subject to two key carbon pricing policies: the European Union Emission Trading System (EU ETS) and the Carbon Price Floor (CPF). The costs of these policies are felt by both businesses and households and so are covered in this subsection, before we consider business- and household-specific policies in subsequent subsections.

The electricity sector is also affected by the Renewables Obligation (RO), which supports generation from renewable sources. From 2014, a number of policies will be introduced as part of the UK Electricity Market Reform, namely Feed-in Tariffs with Contract for Difference (CfD FITs, or simply CfDs), the Capacity Mechanism and the Emission Performance Standards. While detailed recommendations on these future policies fall outside the scope of the study, a short analysis building on existing evidence is provided.

EU ETS

The EU ETS is a cap-and-trade scheme for direct emissions from energy-intensive facilities. The scheme sets a 'cap' on the total emission of certain GHGs (mostly CO_2) that participating organisations may emit. European Union Allowances (EUAs) are created, one for each tonne of CO_2 (or its equivalent, CO_2e, for other GHGs). Participants must surrender one allowance for each tonne of CO_2e they emit. The scheme allows companies to trade emission allowances and thereby determine how and where they reduce emissions.

EUAs are allocated via a mixture of free allocation ('grandfathering') and auctions. Under Phases 1 and 2 of the EU ETS, running from 2005 to 2012, allowances were allocated to industrial operators on the basis of national emissions caps (National Allocation Plans) agreed between the European Commission and individual member states. The majority of these allowances were allocated for free.

In Phase 3, running from 2013 to 2020, the emissions cap is set at an EU level. The cap was 2.04 billion tonnes of CO_2e in 2013, and is to be reduced by 1.74% (37 mt CO_2e) each year. There is also a shift towards the auctioning of allowances in place of free allocation. In 2013 more than 40% of allowances were auctioned, and this share is meant to rise over time. The aviation sector has been included since 2012 for flights fully within the European Economic Area (EEA)[39]. It is subject to a separate cap of 210 $MtCO_2$ per year, which remains fixed until 2020.

In the UK, the EU ETS covered about 230 million tonnes of CO_2e in 2012 (DECC 2013c), some 40% of total UK emissions (Department of Energy and Climate Change, 2013d). In the UK, as well as in the rest of the European Union, companies considered at risk of carbon leakage are exempt from permit auctioning The UK has also proposed compensation for indirect ETS and CPF costs to energy-intensive firms on the basis of their trade and carbon intensity (Department for Business, Innovation and Skills and Department of Energy and Climate Change, 2012).[40]

The price of EUAs has been volatile and positively correlated with the wholesale prices of fossil fuels, reflecting variations in energy demand and the scope for switching commercial energy supplies among sources (Mansanet-Bataller, Pardo and Valor, 2007; Geman, 2005; Chevallier, 2011).

Figure 4.1: EUA price (daily prices of EUA futures with maturity December 2013)

Source: Based on ICE ECX

---

[39] Flights which either begin or end in the EEA, but not both, were also due to be covered by the EU ETS. However, they have received a temporary exemption, pending the outcome of negotiations with the International Civil Aviation Organisation (ICAO).

[40] Companies eligible for compensation should belong to those sectors considered to be exposed to a significant risk of carbon leakage under the EU ETS. According to Directive 2009/29/EC these are sectors or subsectors whose intensity of trade with non-EU countries (defined as the ratio between the total value of non-EU exports plus non-EU imports and the total market size for the Community, i.e. annual turn-over plus total non-EU imports) is above 10 %, and whose sum of indirect additional costs induced by the implementation of the ETS Directive would lead to an increase in production costs of at least 5 % of gross value added (GVA). The UK government proposes to apply an additional filter – that companies applying for compensation demonstrate that their carbon cost (EU ETS and CPF) in 2020 will amount to 5 % of their GVA. Additional sectors may be considered for compensation of the indirect cost of the carbon price support mechanism.

The recent economic downturn saw a contraction in industrial activities, leading to an unexpected reduction in emissions from the traded sectors – with verified emissions decreasing by around 14% between 2007 and 2012 (European Environment Agency, 2013). This in turn contributed to a depression in the price of the EUAs, which plunged from around €29 in April 2008 to less than €4 in April 2013 (Figure 4.1). Despite ongoing discussions on the possibility of backloading some of the allowances (postponing the auctioning of some permits, so that they are removed from the market in the short term, and instead reintroduced at a later stage), it is unclear whether this will be sufficient to raise prices. Analysts have argued that backloading may merely delay, rather than resolve, the oversupply problem afflicting the EU ETS market (see, for example, Gruell and Taschini, 2012; Hitzemann and Uhrig-Homburg, 2013).

Bowen and Rydge (2011) noted that, whilst price volatility is not unusual in cap-and-trade schemes, such volatility can discourage investment, especially in risky and long-term abatement options.

As for the effectiveness of the EU ETS in curbing emissions, early analysis by Ellerman and Buchner (2008) claimed that the scheme led to an abatement of around 50–100 million tonnes of CO_2 per year in its first two years. A more recent review by Martin, Muûls and Wagner (2012) highlighted that the EU ETS may have led to emission abatement in the power sector, but the evidence for wider emission reductions in participating firms is not conclusive. The impact of ETS on innovation and on firms' economic performance was also uncertain.

There is uncertainty about the extent of carbon leakage. *Ex post* analysis of the practical experience so far is fairly limited, but there is general agreement among existing studies that there is no robust evidence the EU ETS has caused substantial carbon leakage to date. For instance, respondents from surveys (Cobb, Kenber and Haugen, 2009) indicate that any impact of the EU ETS on competitiveness has been swamped by other economic effects, such as energy prices, raw material prices or changing international market structures. Econometric analysis of the trade flow of refineries before and after implementation of the EU ETS found no significant changes (Lacombe, 2008). Analysis of the power, cement and iron and steel sectors under the EU ETS for the period 2001–9 found an impact on material costs and turnover due to fuel switching, but little evidence for leakage (Chan, Li and Zhang, 2012). This is also in line with results from Quirion and Demailly (2008) and Anger and Oberndorfer (2008). Bassi, Dechezleprêtre and Fankhauser, 2013 also suggest that the EU ETS has not affected the competitiveness of regulated companies. Abrell, Faye and Zachmann (2011) analyse data from 2005–8 and conclude that being subject to the ETS did not a significantly affect profits and added value during Phase I and the beginning of Phase II, but they find a small negative effect on employment. De Bruyn, Markowska and Nelissen (2010) suggest that some sectors have even obtained windfall profits by passing through the costs of freely allocated emission allowances (as opportunity costs).

In contrast, *ex ante* simulation studies, which estimate the impact of simulated carbon prices, suggest that leakage rates could be fairly substantial, although there are differences across models. In particular, studies are sensitive to assumptions on the substitutability between traded good produced in regulated economies and those produced in unregulated economies (the so-called Armington elasticity of substitution). Where the Armington elasticity is high, carbon leakage effects are high and vice versa. Monjon and Quirion (2009), for instance, find that, under full auctioning of carbon allowances, a high Armington elasticity leads to carbon leakage of 11.4%, versus 4.5% in case of low elasticities. Ritz (2009) also notes that, when firms are trading close substitutes (as is often the case in energy-intensive industries) and compete on price, leakage rates tend to be higher.

Various *ex ante* theoretical studies have also estimated the carbon leakage rate for Phase III of the EU ETS, with estimates ranging from 0 to 39% (Varma et al., 2012; Grubb and Counsell, 2009). Several studies find that the number of sectors at risk of carbon leakage determined by the European Commission for 2013–4 is overstated[41]. The number of sectors, for instance, would reduce significantly should the cost pass-through be taken into account (McKinsey and Company, and Ecofys, 2006; Graichen et al, 2008; Hourcade et al, 2007).

Carbon Price Floor

The Carbon Price Floor (CPF) was introduced in the UK in April 2013 to strengthen the decarbonisation incentives provided by the EU ETS carbon price. The CPF sets a target price which acts as a 'floor'; a minimum carbon price to be paid by power generators subject to the EU ETS. The CPF trajectory started from £16/tCO_2 in 2013 and increases to £70/tCO_2 in 2030 (2009 prices), consistent with the level needed to meet UK and global emission reduction targets (Department of Energy and Climate Change, 2009). In 2020 the CPF is meant to reach £30/tCO_2 (2009 prices).

The CPF is implemented through the imposition of the Carbon Price Support Rate (CPSR) and fuel duty on the purchase of fossil fuels used to generate electricity[42]. The CPSR is calculated as the difference between the CPF target price for a particular year, and an estimate of the ETS price, based on two-year-ahead ETS futures prices. This aims to ensure that the sum of the actual ETS price and the levy paid through the CSPR is close to the estimated price floor. The CPSR for 2013–4 and for 2014–5 is, respectively, £4.94/tCO_2 and £9.55/tCO_2.

---

[41] These are 151 sectors. This represents around 95 % of the industrial emissions under the EU ETS (De Bruyn, Nelissen & Koopman, 2013)

[42] A CCL rate is applied to gas, LPG, coal and other solid fuels used for electricity production, while a fuel duty rate is applied to oils.

By design the CPSR creates a wedge between the carbon price faced by the UK power sector and that faced by the power sector in the rest of Europe. When the price of EUAs is close to the CPF, this wedge will be relatively small. However if, as now, the EUA price remains low, the high CPSR needed to achieve the target floor price could intensify issues of competitiveness and carbon leakage (see simulation evidence cited above). Leakage will occur since the fall in UK demand for EUAs will increase the number of ETS allowances available for use outside the UK. Ultimately, more stringent emission reduction targets within the EU ETS will be crucial for the acceptability of the UK CPF.

A further problem is that, since the rates are announced two years in advance, to achieve a stable carbon price at the target they require accurate forecasts of future ETS prices. The evidence so far suggests that futures prices do not provide a sufficient guide. In 2011, a EUA price of £14.21/tCO_2 was forecast for 2013 (HM Treasury, 2011a). This forecast far exceeded the actual spot rice of below £5/tCO_2 in 2013. This resulted in a lower CSPR than required to reach the CPF target, and meant that the combined carbon price from the sum of the EU ETS price and CPSR fell well below the target CPF of £19.16 (in 2013–4 prices; this is equivalent to £16/tCO_2 in 2009 prices; HM Treasury, 2011a). Such uncertainty is highly damaging to the CPF, since its key aim is to keep the carbon price stable near the desired level, if the EUA price is below the target level.[43]

Assuming that the market can forecast carbon prices more accurately over a shorter horizon, this discrepancy could be reduced by setting the CPSR with a shorter lag, perhaps one year ahead rather than two. However, generators hedge against future price uncertainty and their hedging activities typically extend well beyond a year. Setting the CPSR nearer to its implementation date might therefore require generators to hedge against uncertainty in future CPSR rates as well, since one year ahead prices are still an imperfect guide. There are likely to be trade-offs involved. The Government should explore with the power sector possible routes to reduce carbon price volatility, including the pros and cons of setting the CPSR closer to the time it is paid.

A further issue for the CPF is that of credibility. As with the fuel duty escalator[44], a CPF trajectory has been announced, stating a target price in future years. However, if wholesale electricity prices were to rise substantially, as happened to oil prices, then one way in which governments could ameliorate the impact on final prices would be to switch to a lower CPF trajectory, if EUA prices are low enough for the CPSR to be positive. If energy users believe that this will happen, they will have less incentive to invest in low-carbon technologies now. This is a difficult issue to tackle, since it depends on the beliefs of households and firms

[43] The CPF does not produce any aggregate abatement, since any domestic reduction in demand for carbon simply leaves more permits available for other countries to use at a given price.

[44] The fuel duty escalator imposed pre-determined above-inflation increases in the main rates of fuel duty in each Budget between March 1993 and November 1999. Originally introduced in March 1993, a real increase of 3% was set for future Budgets. This rose to 5% in the November 1993 Budget, and further increased to 6% in the July 1997 Budget, before being abandoned in November 1999. For more information, see Johnson, Leicester and Stoye (2012).

about how a government will choose to act, but it is nevertheless important in understanding whether policy will be effective.

Renewables Obligation and Contracts for Difference

Large-scale generation of electricity from renewable sources is supported by the Renewable Obligation (RO). Introduced in 2002, the RO sets a quantitative requirement on electricity suppliers to purchase a given proportion of electricity from generators using renewable sources, through a system of tradable RO certificates (ROCs). A 'buyout' price per unit has to be paid by suppliers who do not meet their RO quota, and the revenue is recycled to suppliers in proportion to their certificates, providing additional incentive to acquire renewable electricity, but also driving the market price of ROCs up.

Since its introduction, the RO has received some criticism, in particular when its performance was compared with price–setting instruments such as Feed-in Tariffs. For instance, Mitchell, Bauknecht and Connor (2006) highlighted that the RO system was less effective at increasing the share of renewables compared with the German system of fixed feed-in tariffs, because they exposed renewables generators to higher price, volume and balancing risk.

Several policy assessments showed that the UK RO produced renewable electricity at a higher cost than a FIT (European Commission, 2005). This was found to hold for wind, biogas and small-scale hydroplants; in each case, the UK paid more for these developments than Germany or Denmark, despite generation costs being comparable in all three countries.

These findings were later confirmed by research by Toke (2007) and by Butler and Neuhoff (2008) on wind power. They suggested that, once the difference in the wind resource has been taken into account, the price paid for wind energy has been lower and deployment has been greater in Germany than in the UK.

Box 4.1: Contract-for-Difference, Capacity Mechanisms and Emission Performance Standards under the Electricity Market Reform

In December 2010 the UK government announced a programme of 'Electricity Market Reform' (EMR) to support low carbon technologies (including, for the first time, nuclear and carbon capture and storage, CCS) and ensure continued energy security. The proposal included three key policies which will come into force in 2014: a system of Feed-In Tariffs with Contract-for-Difference (CfD), a new Capacity Mechanism, and an Emission Performance Standard (EPS) for new power stations. These measures will represent a significant change to the present policy landscape and they have received a fair amount of comments and criticism. While it is too early to assess their effectiveness, as some of their specifications still need to be defined, this box summarises the main features of EMR.

CfDs are long-term contracts between an electricity generator and a government-owned counterparty, enabling the generator to stabilise its revenues at a pre-agreed 'strike price' for the duration of the contract. When the market price for electricity (the 'reference price') is below the strike price, generators with CfDs receive a payment to make up for the difference, while, when the reference price is above the strike price, the generator pays back the difference to the counterparty. Strike prices will initially be set by the government, but will gradually be replaced by auctioning – possibly as soon as from 2017 for some technologies. The budget for CfD will be capped under the Levy Control Framework

The *Capacity Mechanism* rewards the provision of electricity capacity necessary to meet an optimal level of capacity margin. This will be determined by the government, following a four-year forecast of future peak demand. The total amount of capacity needed to ensure security of supply will be contracted through a competitive central auction. Successful bidders will enter into capacity agreements, committing to provide electricity when needed in the delivery year(s) in return for a steady payment, or face penalties. The costs of the capacity payments will be shared between electricity suppliers in the delivery year.

The *Emissions Performance Standard (EPS)* will set an annual limit on the total amount of CO_2 per unit of installed capacity that new fossil fuel power stations are allowed to emit. The EPS will initially be set at a level equivalent to 450 gCO_2/kWh (at baseload) for all new fossil fuel plant, except Carbon Capture and Storage (CCS) demonstration plants. The level of the EPS on the date of consent of a new power station will apply for the economic life of the installation (grandfathering).

Source: Department of Energy and Climate Change (2012c).

The higher support costs in the UK were explained by the higher risks involved for developers and the high ROC price (Resch et al., 2006). The need to secure contracts in a competitive market meant that the RO created the following risks (Mitchell and Connor, 2004): price risk, volume risk and market risk. Price risk is uncertainty over the future price, which is not known beyond the short-term contract and subject to fluctuation. Volume risk is the uncertainty about the quantity of power that will be needed in the future. Market risk is uncertainty about the generation value caused by potential future variation in market rules.

In response to some of this criticism, the RO underwent a number of changes. The most significant was the introduction, in 2009, of varying rates of certificate allocation across technologies, in order to provide a greater incentive to renewable sources further from the market but with the potential to deploy at a large scale (Bowen and Rydge, 2011).

While these reforms have been effective in addressing some of the problems, not all of them have yet been eliminated. Woodman and Mitchell (2011) stressed that, even under the redesigned RO, generators have no guaranteed market, as they are still required to negotiate a price for their output with suppliers. The resulting uncertainty about future revenue penalises in particular small independent developers and new entrants. Pollitt (2010) has also suggested that the element of the obligation system by which the payments for suppliers 'buying out' their obligations are recycled into the system should be dropped, as it leads to higher ROCs market prices hence contributing to higher cost per kWh to end users.

Some of these issues will be addressed by the new *Feed-In Tariffs with Contract for Difference* (CfD FITs, or simply CfDs) devised under the UK Electricity Market Reform (see Box 4.2). The RO will be closed to new generation from April 2017, with a transition phase between April 2014 and March 2017 during which new renewable generating stations will be able to choose between support under the RO or under CfD. The CFDs could potentially reduce some of the uncertainties surrounding the RO, as they aim to stabilise the prices received by low carbon generation around a 'strike price'. As long as strike prices are set centrally by the government, however, they will remain exposed to the same criticism received by the RO banding, including the risk of asymmetric information and lack of transparency. It will be important, therefore, that auctioning is introduced swiftly. Adequate support measures to small generators will also have to be devised, in order to ensure they have access to the new CfD market.

The CFDs and other key policies proposed under the Electricity Market Reform are briefly discussed in Box 4.1.

4.2 Policies focused on business energy use

Currently the UK business sector is subject to a number of climate change related policies which apply, directly or indirectly, a price on carbon. Key policies are the Climate Change Levy (CCL), the Climate Change Agreement (CCA), the Carbon Reduction Committment Energy Efficiency Scheme (CRC)[45], as well as the European Union Emission Trading System (EU ETS) - the latter is described in Chapter 4.1.

Carbon is explicitly priced by the EU ETS and the CRC, whose rates are expressed in terms of pounds sterling per tonne of CO_2.[46] In contrast, carbon is 'implicitly' priced through the CCL and the CCA. These are imposed on energy use, with tax rates expressed in terms of physical units (e.g. pence/kWh of electricity or pence/kg of coal).

[45] Formerly the Carbon Reduction Commitment. The scheme was renamed in 2010.

[46] Although in principle the EU ETS and CRC both provide explicit prices for carbon, they are applied in practice by charging for use of inputs, with a pre-specified conversion rate from input fuel to carbon.

Box 4.2: A history of the Carbon Reduction Commitment Energy Efficiency Scheme

The CRC has been the subject of almost constant revision since its announcement and now looks significantly different from its initial design.

The original proposal, announced in 2007 (Department for Trade and Industry, 2007), was intended to result in a mandatory cap and trade scheme covering emissions by medium and large firms and public sector organisations which were not already participating in the EU ETS directly. In a first phase of the scheme emissions permits were meant to be sold at a fixed price, with firms required to purchase their expected needs for the year ahead. In a second phase permits were meant to be auctioned. Firms could also buy permits at a price determined by the higher of a fixed 'safety valve' or the EU ETS permit price. CRC participants were required to monitor their energy use, and report their energy supplies annually in a Footprint Report during the first year of each phase, and an Annual Report after the end of each compliance year. Revenues were to be recycled to firms, depending on their emissions and on how each firm performed on a 'league table'. Ranking on the league table was planned to depend on three measures: the percentage reduction in the level of firms' emissions, the reduction in emissions relative to turnover and energy-efficiency actions taken before the CRC was implemented.

The policy was implemented in April 2010. Since its introduction, however, several stakeholders have argued that the scheme was overly complex and administratively burdensome (Department of Energy and Climate Change, 2013e). Consequently, in August 2010, the government expressed its intention to simplify the CRC. In October 2010 it was announced that the revenue recycling element of the scheme was to be removed and the timing of the scheme was changed: firms would be able to pay for the emissions they generated at the end of the year rather than their expected emissions at the start of the year (HM Treasury, 2010).

A range of simplification measures were then announced in November 2010 (Huhne, 2010) and in June 2011 (Barker, 2011) and a formal consultation published in March 2012 (Department of Energy and Climate Change, 2012d). The proposed reforms included delaying the start of the trading phase by two years. The number of fuels covered by the scheme was reduced from 29 to just 4: electricity, gas, kerosene and diesel for heating. The comprehensive reporting required by firms was simplified. It was also proposed that auctioning be entirely replaced by a fixed price uncapped allowance sale, with subsequent trading in secondary markets. This was accompanied by a plan to require firms to buy allowances at the start of the year (as intended in the initial CRC proposals and subsequently revised), with any unmet obligation bought out at a higher price at the end of the year.

In December 2012 it was announced that the league table would be abolished from 2013 (HM Treasury, 2012). The data for the table would still be collected and published, but organisations would no longer be ranked according to their performance. It was also announced that a full review of the CRC effectiveness would be held in 2016 and that 'the tax will be a high priority for removal' (HM Treasury, 2012). The following week the response to the consultation was finally published (Department of Energy and Climate Change, 2012e). This reaffirmed the proposed changes, and went further on reducing the number of fuels to only two: electricity, and gas for heating. The revised scheme came into force in May 2013.

Support for small-scale renewable technologies by individual firms is mainly provided through Feed-In Tariffs (FITs). Since these also apply to households, we defer their discussion until Chapter 5.4.

The *CRC* taxes the energy consumption of large public and private sector organisations which are not already covered by the EU ETS and for which less than a quarter of emissions are covered by a CCA. When introduced, in 2010, it covered 29 energy sources. These were narrowed down to only two – electricity and gas – after simplifications introduced in 2012. The tax rate has been fixed at £12 per tonne of CO_2 since the beginning of the scheme.

A key issue for the CRC scheme has been administrative complications and costs. While the government has proposed several amendments aiming to simplify some of these issues, the scheme remains cumbersome. Furthermore, the complex history of the instrument, which led to a high number of revisions (see Box 4.2), has increased uncertainty on the near-term design of the instrument, as well as on its long-term existence, discouraging investment in energy conservation and low carbon technology.

Like the CRC, the *CCL* applies a tax downstream on energy use in the business sector (rather than upstream on primary energy providers). The CCL targets electricity, gas, LPG and solid fuels (mostly coal).

According to Martin, de Preux and Wagner (2011), the CCL has reduced business energy intensity, especially at larger and more energy-intensive plants. This effect was mainly driven by a reduction in electricity use, which has fed into a reduction in CO_2 emissions from businesses.

Despite this contribution to CO_2 mitigation in the UK, some flaws in the design of the CCL remain. Notably, as discussed in Chapter 6, the implicit carbon prices on energy from different sources differ substantially. Given its carbon content, coal is relatively lightly taxed compared with gas and electricity, with some ascribing this to the desire to put less of a burden on the coal industry than other energy providers (Pearce, 2005; Helm, 2010). The carbon price charged for electricity is the highest, and does not distinguish between low and high carbon energy sources used by the suppliers. As a result, implicit carbon prices vary across firms and sources in a way that makes abatement less effective and more costly (see Chapter 6). It may also be noted that the CCL was levied downstream on business, rather than upstream on electricity generators, essentially to protect households from higher electricity prices. However, costs still inevitably fall on households, as consumers of business products whose prices rise, suppliers of labour to firms whose wages fall or owners of firms whose profit shares fall.

The *CCA* is essentially a discount on the CCL granted to energy-intensive firms which commit to given energy or emission targets agreed with the government. This was meant to reduce potential detrimental effects of the CCL on the competitiveness of these firms. Early studies have found that the CCA scheme contributed to significant carbon savings (see, for example,, AEAT, 2004; Barker, Ekins and Foxon, 2007; Ekins and Etheridge, 2006). In particular, Ekins and Etheridge (2006) suggested that, even though the targets imposed by CCAs on plants were met relatively easily, the agreements contributed to increased energy savings through their 'awareness effect'. Firms themselves have claimed that the CCA were effective in attracting managerial attention to energy efficiency (Environmental Audit Committee, 2008).

Others, however, have argued that CCAs have not been very demanding, given the way targets were negotiated and the underlying trend in energy efficiency improvements (Bowen and Rydge, 2011). Martin, de Preux and Wagner (2011) suggest that, between 1999 and 2004, negotiated CCA targets were unlikely to have placed binding constraints on energy use by participating companies, and their impact on energy intensity has been much lower than the CCLs. They estimate that, had the CCL been applied to all plants without rebates, it would have decreased aggregate energy expenditures in manufacturing by at least 5% and aggregate electricity consumption by at least 12%. The same study also found no evidence that the full CCL rates had an impact on firms' employment, gross output or indeed the likelihood that a firm would exit the market, compared with the CCA. It appears therefore that the CCL did not cause firms to shed jobs or lose revenue relative to CCA firms.

These studies take into account CCAs tax discounts of 80%, and data up to 2004–5. Although today the tax differential between the CCA and CCL is slightly different (in April 2011 the discount for gas, coal and LPG was decreased to 65%, while the electricity discount was raised to 90% from April 2013), it is apparent that tax discounts lead to less innovation and worse energy performance than a full application of the CCL, and that the impacts on competitiveness and employment of the CCL have so far been negligible. Bassi, Dechezleprêtre and Fankhauser (2013) use firm-level data up to 2010 to confirm the results from Martin, de Preux and Wagner (2011) relating employment and output. This again suggests that the CCA may still not be justified on the basis of competitiveness alone.

5. Current UK household energy use policies

Arun Advani, Paul Johnson, Andrew Leicester and George Stoye

5.1 Policies affecting domestic energy prices

As is the case for businesses, the cost of electricity used by households is increased by the range of policies discussed in Chapter 4 which increase the costs of generation (the ETS, CPF and large-scale renewable support). Unlike businesses, however, there are no policies which raise household energy prices specifically at the point of consumption based either on the quantity of energy consumed or its assumed carbon content: there is no domestic equivalent of the CCL/CCA or the CRC, for example. This means there are no policies at all which raise the price of gas or non-metered fuels used by households. In addition, households also pay a reduced rate (5%) of VAT on energy, which acts effectively as a subsidy to domestic energy consumption of around 14.3%. Among EU countries, the UK charges the lowest VAT rate on domestic energy. Most other member states tax domestic energy at the full rate of VAT, though a few others have reduced rates (13.5% in Ireland compared with a standard rate of 23%, 13% on gas in Greece compared with a standard 23%, 10% in Italy for some units of domestic electricity compared with a standard rate of 21%, 5% in Malta for electricity compared with a standard rate of 18%, 6% in Luxembourg compared with a 15% standard rate).[47]

It is clear, therefore, that policy towards domestic energy use has focused much more on encouraging improvements in energy efficiency (CERT, CESP, ECO and the Green Deal) and small-scale low-carbon generation (FITs) together with targeted support for energy bills (WFP, WHD and CWP). This reflects concerns about the equity implications of higher energy prices and stated objectives to reduce fuel poverty (see Chapter 2).

However, it is worth being clear that many of these policies (CERT, CESP, ECO, FITs and WHD) are delivered through energy companies who are able to recoup the costs through levies on domestic energy bills. It is not immediately clear precisely how this is done, whether through higher prices per unit of energy or a fixed levy as part of the standing charge, for example. Relative to policies which price energy use or generation directly, the precise impact of these supplier-delivered levies on individual bills is much harder to calculate, and so the effect on equity and fuel poverty is much less transparent.

[47] See http://ec.europa.eu/taxation_customs/resources/documents/taxation/vat/how_vat_works/rates/vat_rates_en.pdf.

As we discuss in more detail in Chapter 6, if we assume these policies increase unit energy costs for households and ignore the VAT subsidy, the combined effect of policies on the price of energy used by households and businesses is quite similar. However, once we add the effective subsidy to households from the reduced rate of VAT, policies increase domestic energy prices by much less than they increase business prices. Indeed, in the case of gas and non-metered fuel, the full effect of policies is to *reduce* domestic prices.

5.2 Policies affecting incentives for small-scale renewables

The *Feed-in Tariffs (FIT)* scheme provides a financial incentive for households, community organisations and commercial and industrial businesses to generate renewable and low-carbon energy. Installations (up to 5 MW) which use the following low-carbon and renewable technologies are eligible for the scheme: hydro, anaerobic digestion, solar photovoltaic (PV), wind, hydro and micro combined heat and power. Under the scheme, large energy suppliers make guaranteed payments to system owners for the electricity they produce. This cost is then recouped through the electricity bills of consumers. Eligible installations enrolled in the scheme receive two types of payments:

1) Under the *'generation tariff'* the installation owner is paid for each kWh of energy generated by the system, regardless of how the energy is used. The level of the generation tariff varies by installation type, size and technology, and has declined since the start of the scheme in line with changes to the design of the regulation.
2) Under the *'export tariff'* additional payment is made for each kWh of electricity exported to the grid by the installation. The level of the export tariff is smaller per kWh than the generation tariff, but has been, and is expected to continue to be, more stable over time.

Since the scheme was launched in April 2010, the rate of FIT installation registration has been high. As of April 2013, almost 380,000 installations have been registered under the scheme in England, Scotland and Wales. In England and Wales this amounts to approximately 1 installation for every 50 households. FIT installations now account for around 2% of total UK generating capacity. The aim under the policy is to increase the number of registered installations to a total of 750,000 by 2020.

During certain periods, the number of registered installations was far greater than expected prior to the launch of the scheme. This is in part due to the rapid falls in the cost of PV installations over the period. Evidence suggests that the cost of a typical installation was 45% lower in 2011 than estimates performed in 2009 had predicted.[48] This meant that at the initial tariff levels the FIT scheme was delivering far more generous rates of return to adopters than had been expected,

48 http://webarchive.nationalarchives.gov.uk/20121217150421/http:/decc.gov.uk/assets/decc/consultations/fits-review/4310-feedintariff-comprehensive-review-phase-1-impact.pdf .

leading to larger take up and higher costs to consumers. Initial estimates were that by 2020 the policy would cost consumers around £560 million (in 2008 prices) per year in terms of subsidy.[49] More recent estimates supporting plans to reduce tariff rates suggested that without any change the actual cost in 2020 would have been £4,970 million (2011 prices), almost nine times the earlier estimate.[50]

As a result, there was an urgent review of tariff rates for PV to reduce the size of the subsidy. In October 2011, DECC announced that PV installations with a reference date from 12 December would be subject to a much lower tariff. Following legal challenges, the reduction eventually occurred in March 2012.

Figure 5.1: Number of domestic PV installation registrations per week, July 2009–July 2013

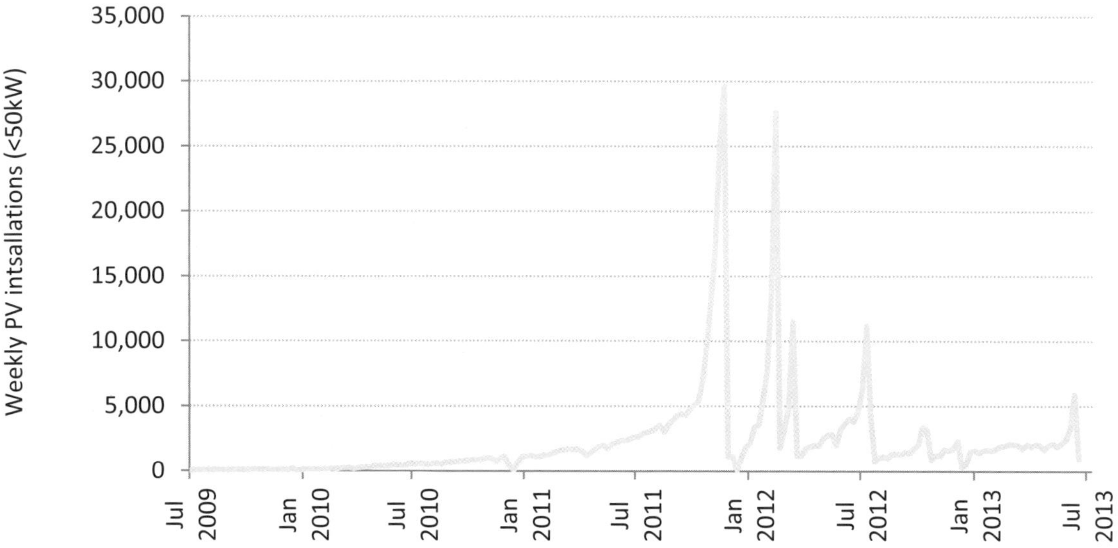

Source: Figures to January 2012 from http://webarchive.nationalarchives.gov.uk/20130109092117/http://www.decc.gov.uk/en/content/cms/statistics/energy_stats/source/fits/fits.aspx, based on date of commission. Later figures from https://www.gov.uk/government/statistical-data-sets/weekly-solar-pv-installation-and-capacity-based-on-registration-date, based on date of registration.

Figure 5.1 shows the number of domestic PV installations that were registered in each week between July 2009 and July 2013. Of note are the peaks, and subsequent falls that occurred in December 2011 and March 2012, corresponding to the planned and actual implementation of the reduced rates. Similar spikes are also observed in July and October 2012, preceding further

[49] http://www.fitariffs.co.uk/library/regulation/090715ImpactAssessment.pdf, Table 1 for 'Lead Scenario'.

[50] http://webarchive.nationalarchives.gov.uk/20121217150421/http:/decc.gov.uk/assets/decc/consultations/fits-review/4310-feedintariff-comprehensive-review-phase-1-impact.pdf, Table 12, 'Central Scenario'.

tariff changes for some installations on 1 August 2012 and 1 November 2012 respectively.

The total cost of the FIT scheme was just under £500 million between April 2012 and March 2013. Even with the reduction in tariffs, the cost to consumers in 2020 is still estimated at around £2,840 million (2011 prices), around five times the initial estimate.[51] Ofgem estimates the total cost of the scheme at £7.9 billion (in 2009 values) cumulative to 2030. In terms of the distributional impact of the FIT scheme, it is important to recognise that the cost of FIT payments made to installation owners by electricity suppliers is passed on to all of the electricity consumers in the service areas of the electricity suppliers. This means that the FIT scheme is paid for by electricity consumers rather than taxpayers. While the FIT framework authorises the electricity suppliers to pass on the costs of FIT payments to electricity consumers, it does not specify how this should be in done in practice. As a result, it is unclear how electricity suppliers recoup these costs, or how this burden is distributed across different household income groups.

The distribution of the ownership of FIT installations is clearer. Grover (2013), suggests that FIT installations are concentrated most densely in areas which are less deprived, and have lower unemployment rates, higher social status and greater rates of outright property ownership.

5.3 Policies supporting domestic energy efficiency

Policies which provide support for household energy efficiency have until recently been delivered largely as obligations on energy suppliers and generators. The Carbon Emissions Reduction Target (CERT) began in 2008, and the Community Energy Savings Programme (CESP) in 2009.[52] Both ran until the end of 2012. These programmes were designed to deliver insulation and other measures to improve the energy efficiency of homes, with CESP explicitly focused on homes in deprived areas and on a whole-house approach. They delivered measures to households that are estimated to save over 313 $MtCO_2$ emissions over their lifetime (broken down as 296.9 million (CERT) and 16.3 million (CESP); see Ofgem 2013c,d). More recently, CERT and CESP have been replaced by the Green Deal and the Energy Companies Obligation (ECO). The Green Deal provides loans to pay for the upfront installation of efficiency measures. These are then repaid through energy bills, which are tied to dwellings rather than individuals. ECO provides direct support through energy suppliers for people living in hard-to-treat properties or deprived areas.

---

51 http://webarchive.nationalarchives.gov.uk/20121217150421/http:/decc.gov.uk/assets/decc/consultations/fits-review/4310-feedintariff-comprehensive-review-phase-1-impact.pdf, Table 15, 'Central Scenario'.

52 CERT and CESP were follow-ups to other obligations on suppliers including the Energy Efficiency Standards of Performance, which ran from 1994 to 2002, and the Energy Efficiency Commitment, which ran from 2002 to 2008. Both supported the installation of insulation measures and were funded through energy bills (Preston and Croft, 2012). Warm Front, which ran from 2001 to 2012, was funded through general taxation and provided free insulation and heating system improvements to poorer households (for more see Leicester, 2006).

The emissions savings targeted under the ECO are much smaller than those achieved under CERT and CESP: 28.8 $MtCO_2$ between January 2013 and March 2015, compared with a combined 313$MtCO_2$ over the period between 2008 and 2012. This is reflected in the fall in predicted new cavity wall insulation from 550,000 in 2012 (the last year of CERT and CESP) to around 398,000 in 2013, and loft insulations from more than 800,000 to under 200,000. This may, in part, reflect the fact that there are fewer homes over time without loft or cavity wall insulation. However, Department of Energy and Climate Change (2012f) estimate that in 2013 there remain around 0.7 million 'easy to treat' uninsulated cavity walls, a further 0.9 million 'easy to treat' cavity walls with limited potential[53], and 5.7 million lofts that could benefit from top-up insulation (again, roughly equal to the number of lofts insulated either professionally or by DIY under CERT).

Part of the expected change simply reflects the incentives under the Green Deal and ECO relative to CERT and CESP. Solid wall insulation will be delivered through ECO (and was expected to increase from around 22,000 installations in 2012 to 42,000 in 2013), whilst cavity wall and loft insulation will be delivered through the Green Deal. In moving from delivering these policies through supplier obligations (where they are heavily subsidised or free and suppliers have to carry out actions to meet a specific goal) to a loan-funded mechanism where there is no explicit target for what is delivered in total, it is perhaps not surprising that these trends were expected. Indeed, there appears to be some initial evidence that cavity wall insulation rates under the Green Deal and ECO are much lower than anticipated (Committee on Climate Change, 2013b), in part because of delays in setting up of the necessary finance mechanisms which has meant Green Deal loans are not yet widely available.[54]

With no explicit target set, emissions savings from the introduction of the Green Deal are uncertain. Savings will depend on the number of households taking it up and the type of measures they install; this will in turn depend on things like the rate of interest charged by Green Deal providers and the cost of Green Deal assessments. No formal trialling of the Green Deal was conducted to examine consumer responsiveness to these prices. However, a precursor scheme called 'Pay as You Save' (PAYS) was trialled in some areas by energy and other companies, which offered interest-free loans and free energy assessments (Department of Energy and Climate Change and Energy Saving Trust, 2011). Around 55% of households who had their homes assessed ultimately signed up to have measures installed. Evidence from the final impact assessment (Department of Energy and Climate Change, 2012g) suggests the government expects interest rates of around 7.5% on average together with an assessment cost estimated at £112.50. This suggests that in financial terms the Green Deal

[53] Properties with 'limited potential' are uninsulated but are likely to already have a relatively high standard of thermal insulation. This includes properties built in England and Wales between 1983 and 1995, and in Scotland between 1984 and 1991 (Department of Energy and Climate Change, 2013f).

[54] See, for example, BBC News website, 27th June 2013, 'Only four people sign up for flagship Green Deal', http://www.bbc.co.uk/news/business-23081896.

package will be less generous than PAYS such that conversion rates between assessments and installation of measures may be considerably lower.

The economic rationale for a policy like the Green Deal is clearest if households fail to install various insulation measures because of upfront credit constraints. Leicester and Stoye (2013) analyse household-level data on different common insulation measures such as thick loft insulation and cavity wall insulation, and find no compelling evidence that low income (or other variables that might proxy those who are credit constrained such as receipt of means-tested benefits, low education or being unemployed) is correlated with a lower probability of having the measures. These results hold not only for relatively cheap measures but also for more expensive measures such as full double glazing (estimated to cost around £4,500). The lack of any obvious effect of income on insulation could reflect the success of past policies delivered through energy suppliers, including CERT and its forerunners (various Energy Efficiency Commitments), which at least partly targeted poorer households and those in social housing. However, analysis carried out on much earlier data from the late 1980s, before such schemes, also found a very limited role for income in explaining the presence of insulation measures (Brechling and Smith, 1992).

A greater barrier to the installation of domestic energy efficiency measures may instead be landlord–tenant relationships, with incentives to install energy efficiency measures in rented properties poorly aligned across the two parties. Leicester and Stoye (2013) found that private renters who had been resident in their property for less than two years were 11% less likely to own loft insulation in 2010 than a comparable owner-occupier. Interestingly, social tenants were no less likely to own loft or cavity wall insulation than owner-occupiers, and were 3% more likely to own full double glazing, even though they were significantly less likely to own such measures in 1986 (Brechling and Smith, 1992). This further suggests that previous policies aimed at supporting domestic energy efficiency, which focused most on poorer social households, have been relatively successful. The Energy Act 2011 includes provisions to ensure that landlords install cost-effective measures under the Green Deal from 1 April 2016, but private renters may find it difficult to take advantage of the policy before this date. Taken together, then, it appears that the Green Deal is addressing a market failure which does not appear to be the key barrier to the installation of efficiency measures, and the high interest costs look like being significant deterrents to take-up, at least relative to previous obligations, which largely delivered measures for free. This may explain why an additional cashback scheme has been introduced, at least for a limited time. Of course, the benefits in terms of information and visibility could also prompt some households to take up new installations, and would therefore boost installation rates.

5.4 Policies supporting domestic energy bills

As summarised in Table 3.1, aside from the reduced rate of VAT there are three main policies that provide support for household energy bills, either directly or through labelled cash payments: the Winter Fuel Payment (WFP), Cold Weather Payments (CWP) and the Warm Home Discount Scheme (WHD).

The *WFP* is the largest. Any household containing someone aged at the female State Pension Age or above is eligible. The payment is automatic, universal and tax free, with a higher payment made for those aged 80 and over. The WFP costs more than £2 billion per year, though expenditure was higher in the two periods between 2004–5 to 2005–6 and 2008–9 to 2010–11 because of a series of supposedly 'one-off' supplements. The changing generosity of the scheme is noted by Advani et al. (2013), who estimate that the WFP was equivalent to around 46% of the energy costs for people aged 60 to 79 in 2005–6, and 76% of costs for those aged 80+. By 2013–4, this is estimated to fall to 13% and 22%, respectively, in part because of rising energy costs and in part because of lower WFP rates. Those proportions would be the lowest since around 1999.

It is hard to understand precisely the rationale for WFP being a universal benefit. Evidence from Beatty et al. (2011) suggests that the labelling of the benefit as a 'fuel' payment and its delivery in the winter months leads recipients to spend far more of it on energy (around 41% on average) than would be expected from the same amount given as a regular cash transfer (the standard 'income effect' for energy, estimated at around 3%). As noted by Crossley and O'Dea (2010):

> *If the aim of the payment is to encourage older individuals (regardless of their income) to increase their fuel consumption, then it seems to be a reasonably successful (albeit expensive) policy.*

It may be that policymakers believe that, because of concerns about costs, some people use less energy than perhaps they 'should' to heat their homes adequately, meaning they risk damaging their health. Department of Energy and Climate Change (2011a) estimate that households in the poorest income group spend on average just 60% of the estimated necessary spend on energy in order to heat their homes. However, it is far from clear that such under-heating is true for all elderly households, or that a labelled cash transfer would be the most effective way to deal with this (relative to a voucher or direct bill subsidy), since people could still choose not to spend the money on energy.

If instead the WFP is aimed at reducing fuel poverty then spending large amounts of money giving additional income to better-off pensioner households is poorly targeted. Given that there are already benefits targeting poorer pensioner households (the Pension Credit) it would be feasible to restrict eligibility for the WFP to those households – Adam, Browne and Johnson (2012) suggest that doing so alongside restricted eligibility for concessionary television licenses would raise around £1.4 billion per year, which they suggest would be almost sufficient to meet the costs of the Dilnot proposals on long-term care. One concern with this would be non-take-up of benefits: Department of Work and Pensions (2012) estimates for 2009–10 suggest that only around 73–80% of those eligible to receive the guarantee credit element of Pension Credit (that targeted most specifically on low income pensioners) actually took it up. Restricting WFP to

Pension Credit recipients would therefore lead to some poorer households losing out.

Even if the WFP were restricted to poorer, older households it may not be the most effective targeting of such support to reduce fuel poverty. Using data from the English Housing Survey 2010–11, for example, we estimate that among households containing someone aged 60–79, fuel poverty rates were 21.3% if nobody received Pension Credit, and 22.7% if they did: a relatively small difference. For those aged 80+, rates were 29.8% and 29.9% respectively: virtually no difference at all. Receipt of other benefits may be as good or even better a marker for fuel poverty: for example, 43% of those on Jobseeker's Allowance were fuel poor, as were 22% of those on Council Tax Benefit and 19% of those on Income Support. This might suggest targeting support at poorer households more generally, rather than poorer, older households specifically.

Further, the definition of fuel poverty we consider here is based on households needing to spend at least 10% of their income to heat their homes adequately. Under the new definition proposed by Hills (2012) and discussed in Chapter 2, households are fuel poor if they have both relatively high energy needs and low incomes. Department of Energy and Climate Change (2013a) suggest that, on this measure, households aged 60+ are at the *lowest* risk of fuel poverty, whereas under the old measure it is households aged 25 to 49 who are least at risk.

CWP are administered through the Social Fund by the Department for Work and Pensions. Each UK postcode is linked to one of 92 national weather stations. A payment of £25 is automatically made to eligible recipients following a period of seven consecutive days (between 1 November and 31 March) when the daily mean temperature at the relevant station is recorded or forecast to be 0° C or below. Eligibility for CWP is determined by receipt of means-tested benefits:

- Recipients of Pension Credit are automatically eligible.
- Recipients of income-based Jobseeker's Allowance or Income Support are eligible if they also receive a disability or pensioner premium, or have a young (under five years old) or disabled child.
- Recipients of income-related Employment and Support Allowance who have had a work capability assessment and go on to receive the support- or work-related component of ESA are eligible. ESA recipients who have not had the assessment are eligible if they also receive a disability or pensioner premium, or have a young or disabled child.

Spending on CWP will depend on weather conditions; the recent peak was 2010–11, when 17.2 million payments were made at a cost of £431 million. There is a much clearer rationale for targeting support for energy bills on poorer and vulnerable households in periods of very cold weather. Beatty, Blow and Crossley (2011) use detailed household expenditure data and find evidence that among older, poorer households in periods of very cold weather (occurring around one winter month in forty), there is a trade-off made between 'heating and eating': these households cut back on food expenditure in order to maintain their energy spending. Linking support to income and weather conditions therefore appears sensible, though the evidence from Beatty, Blow and Crossley (2011) may suggest even greater scope to support heating costs during the very coldest periods.

The *WHD* gives electricity bill rebates (worth £135 in 2013–4) to low income and vulnerable households.[55] Energy companies with at least 250,000 domestic customers are obligated to take part, and the costs are recouped through higher energy bills for non-recipients. There are two groups eligible for the WHD rebate:

1) A ***core group*** of low income pensioners. There are two determinants of eligibility for the core group:
    a. Customers aged under 75 who receive just the Guarantee Credit element of the Pension Credit, but not the Savings Credit element (for a single pensioner in 2013-14 this amounts to a weekly income of less than £115.30);
    b. Customers aged 75 and over who receive the Guarantee Credit element of the Pension credit are eligible irrespective of whether they also receive the Savings Credit element (in 2013–4 for a single pensioner this amounts to a weekly income of less than £145.40).[56]
2) A ***broader group*** of other customers are also eligible for a rebate. The criteria are at the discretion of individual energy companies (subject to Ofgem approval); in practice, eligibility is largely determined by receipt of an income-related benefit (Income Support, Income-Related Employment and Support Allowance or Income-Based Jobseekers' Allowance) together with having young children, older people or disabled people in the household; or by receipt of the Pension Credit for those not already part of the core group.[57] These are essentially the eligibility criteria for CWP.

Unlike WFP and CWP, the WHD is paid as a bill subsidy directly rather than a cash payment, though as discussed the labelling effect blurs the distinction between the two somewhat. In principle, assuming that the subsidy is not larger than the total bill, it should make no difference whether households receive money off a bill or receive a cash payment: final disposable income should be the same in either case. However, subsidies (or vouchers) have been shown to have particular effects on behaviour, leading people to increase consumption of the subsidised product beyond what would be expected from a straightforward income effect (Abeler and Marklein, 2010). This suggests that the WHD will see recipients increase their energy consumption. The label 'warm home' might also spur recipients to consume more energy rather than simply treating the rebate as additional cash.

---

[55] There are other aspects to the WHD scheme, including spending by energy companies on social tariffs and some activities carried out by energy companies to help people reduce bills. For details see Appendix B.

[56] Note that the age threshold at which receipt of the Savings Credit element is ignored is set to fall to 65 in 2014–15. The weekly incomes determining eligibility for the different components of Pension Credit vary for couples and for people with particular housing costs or caring circumstances; for more information see Browne and Hood (2012).

[57] The eligibility criteria for the broader group for British Gas, for example, in 2012-13 can be found at http://www.britishgas.co.uk/products-and-services/gas-and-electricity/the-warm-home-discount.html.

In terms of the effect on fuel poverty, spending money on bill support is more effective (at least given the '10%' definition of fuel poverty): for a household with an income of £9,000 and a necessary energy bill of £1,000, it would take an income supplement of £1,000 to remove them from fuel poverty compared with a bill rebate of just £100. Department of Energy and Climate Change (2011a) estimated that the WHD scheme would reduce fuel poverty by around 58,000 households in 2014–5.

It is hard to see a clear rationale for having eligibility for the broader group vary across energy companies, unless it is thought they have private information about those among their customer base who are likely to be fuel poor, which is not known to policymakers. Given that energy companies are unlikely to know much about the wider income circumstances of their customers, this seems implausible. Varying eligibility could also reduce the incentive for customers to switch between suppliers if they are uncertain about whether they would be eligible with a different company.

A key difference between the core and broader groups, aside from eligibility criteria, is that those in the core group should automatically receive the rebate: DWP benefit payment records are matched to energy company customer records. Households in the broader group have to apply for the rebate themselves. In their first year report on the scheme, Ofgem (2012a) found that around 600,000 of an estimated 800,000 core group households received an automatic rebate with a further 100,000 subsequently receiving a rebate through a mop-up process. More than 234,000 broader group rebates were paid, though only 42% of the broader group applicants whose application was audited by energy companies could produce evidence of their eligibility. This suggests that many ineligible households received a WHD rebate.

Part of the WHD scheme is to see existing low-price ('social') tariffs for poorer households eventually wound up and replaced with the direct bill subsidy – indeed, Ofgem (2012a) evidence suggests this is being done more quickly than envisaged. Those likely to gain from a switch from a price discount to a fixed bill discount will be people who use relatively little energy and thus have low bills. This will include households already living in relatively efficient homes, who are less likely to be fuel poor.

It is striking that the core group for WHD, who should largely receive automatic rebates rather than having to apply, are older, poorer households who also receive the WFP and CWP. This adds up to a substantial package of bill support, or policies which look very close to bill support, targeted on a specific group of poorer pensioners. Other groups for whom fuel poverty is likely to be a significant issue, such as poorer households with children, disabled people or unemployed people, will be entitled to a substantially less generous package of measures through CWP payments and non-automatic WHD payments.

Figure 5.2, replicated from Advani et al. (2013), shows estimates of the value of total WFP, WHD and CWP eligibility in 2013–4 relative to energy spending by household type and broad position in the total spending distribution (divided into the poorest, middle and richest third). Lone parents are eligible for bill support worth on average £40 per year, or 3.2% of their average fuel expenditure. For lone parents in the bottom third of the spending distribution, the figures is £58 (5.8% of fuel spending). Single pensioners, in contrast, are eligible for support worth £273 per year or 24% of fuel spending. For those in the bottom third of the distribution, support is worth £290 per year (32% of fuel spending), but even for those in the richest third support is worth £238 per year (16% of fuel spend).

Figure 5.2: Average value of bill support eligibility relative to energy spending, by household type and total expenditure tertile, 2013–4 values

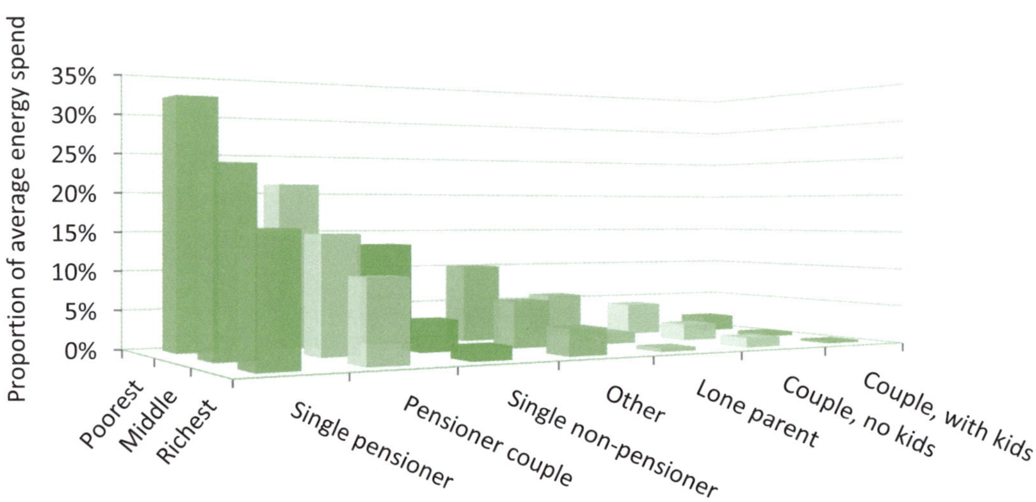

Source: Authors' calculations from 2010–11 Living Costs and Food Surveys. Notes: Based on 2013–4 eligibility criteria and policy values, fuel spending uprated to 2013–4 values using relevant RPI indices. Assumes WHD eligibility where the head of household or their spouse is eligible for the CWP. Assumes eligible benefit units receive one CWP. Data are weighted for household non-response. Excludes Northern Ireland and households who spend less than £1,000 per year on all goods and services.

In addition, these figures are based on *eligibility* for the various policies. For pensioner groups, receipt is automatic; for non-pensioner groups eligible for the WHD, receipt is not automatic. Advani et al. (2013) suggest that many households who are likely to meet the eligibility criteria for the broader WHD group appear to have missed out: around 3.5% of households received a WHD rebate in 2011–12 against around 11.5% who were eligible.[58]

[58] If all eligible households actually received a payment, total spending on rebates would exceed the WHD budget set out in the Levy Control Framework (see Box 3.1).

6. Implications of policies for carbon prices on energy use

Arun Advani, Paul Johnson, Andrew Leicester and George Stoye

Chapter 4 and Appendix A outline the main policies which affect energy prices at different stages of the energy production and consumption process. In this section, we assess what these different policies imply for the effective carbon prices imposed on different fuels used by different end users.

We calculate how policies affect marginal energy prices for groups of consumers and, using information on the carbon content of different energy types, convert the energy price impact into a carbon price. As we described in Chapter 1, carbon is the externality that should be priced.[59] Efficient policies which gave incentives to reduce carbon use at the lowest cost would ideally lead to similar carbon prices across fuels and users.

Our main results are based on 2013–4 values, though at the end of this chapter we estimate how carbon prices may change by 2020 based on current estimates of the policy and pricing landscape in that year. Full details of the methodology are given in Appendix B.

6.1 End users and fuels

We estimate carbon prices for four types of end user:[60]

- **Households**
- **Small businesses,** which are assumed not to face the CRC for their energy and gas use.
- **Medium businesses,** which are assumed to face the CRC.
- **Large energy-intensive businesses,** which participate in the EU ETS directly, and so do not face the CRC, but qualify for a CCA discount on their CCL liability.

We calculate prices for four different fuels: electricity, natural gas, coal and liquefied petroleum gas (LPG). Figure 6.1 displays the overall carbon emissions associated with the use of each fuel, by end user. This chapter focuses on

[59] In fact we also include other GHG emissions within our 'carbon price', as all six of the GHGs identified in Box 2.1 produce an external cost. All prices presented here should be strictly interpreted as 'carbon dioxide equivalent' (CO_2e) prices. Results for the implicit carbon prices associated with electricity and gas use change little when using CO_2 instead of CO_2e, as carbon dioxide accounts for at least 98% of the total GHG content of these fuels (DEFRA and DECC, 2012).

[60] These users are illustrative of the different carbon pricing regimes affecting broad groups, but are not fully exhaustive in terms of the variation in effective carbon prices faced economy-wide. For example, some firms will be eligible for a discounted rate of CCL having signed a CCA but will also face the CRC. Even within the groups we identify there will be variation in prices across individual users: to give two examples, until November 2013, firms in Northern Ireland face a reduced rate of the CCL, and some households may still benefit from reduced rate 'social tariffs' which may be seen as an effective subsidy to their energy use. Nor do we try to calculate the effective carbon prices imposed by non-economic instruments such as environmental regulations that might be incident on different sectors.

electricity and gas. Results for non-metered fuels are presented in more detail in Appendix B.

Figure 6.1: Fuel consumption by end user in 2012

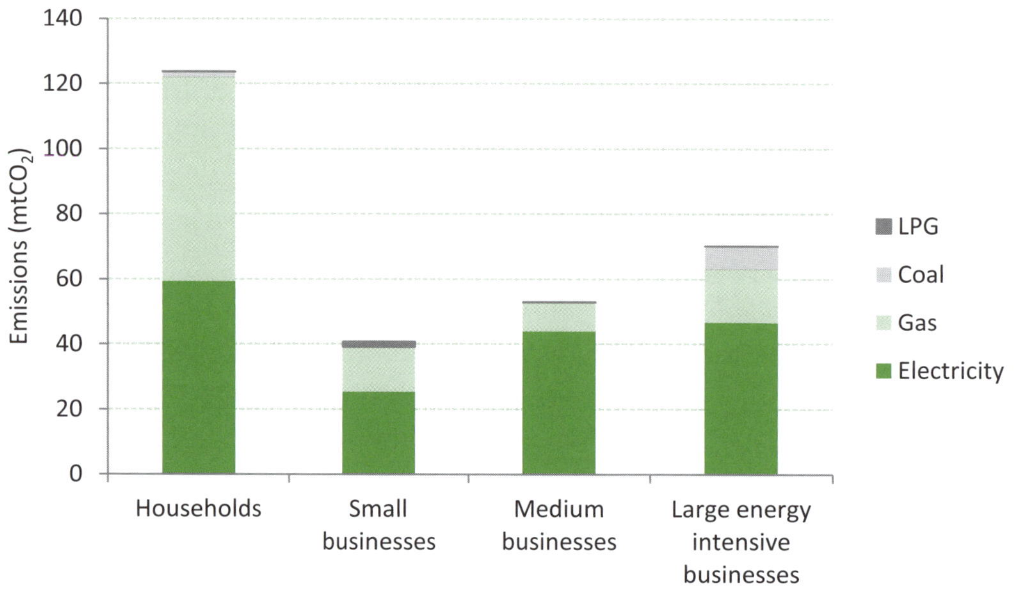

Source: Authors' calculations, based on Department of Energy and Climate Change (2013f)

Box 6.1: The carbon content of electricity: marginal vs. average emissions

When modelling the carbon value of electricity taxes, an important issue is how we determine the carbon content of a unit of electricity.

One quite natural possibility is to use the average carbon intensity of a unit of electricity (total carbon emissions from electricity generation divided by total generation), which depends on the mix of fuels used to generate electricity in a given period. Whilst this is a relatively straightforward concept, it may not be the most relevant number for comparing implicit carbon prices across fuels.

From a carbon pricing perspective, efficient pricing should equalise the incentives to reduce emissions by setting the same price at the margin for a unit of carbon abatement. In other words, if we are thinking about a policy which raises the price of electricity, we want to know how much carbon has been saved as a result. For a sufficiently small but sustained change, this is the carbon content of electricity generated by the long-run marginal plant. At the moment this is predominantly combined-cycle gas turbine (CCGT), since it is relatively cheap and quick to build (Department of Energy and Climate Change and HM Treasury, 2013a). If instead the reduction in capacity came equally from baseload, mid-merit and peaking plant, then the 'grid average' emissions would be the relevant factor to use.

The carbon content of long-run marginal emissions is 0.368 $tCO_2e/12MWh$ (Department of Energy and Climate Change and HM Treasury, 2013b, Table 1). This is lower than the average 'grid mix' carbon content, which is 0.494 tCO_2e/MWh (Department of Energy and Climate Change and HM Treasury, 2013b, Table 1), since the average also includes more carbon-intensive coal-fired power along with CCGT. By 2020 the average is projected to fall significantly, and be lower than the long-run marginal. This is because more of electricity demand will be met by emissions-free nuclear and renewables, and less by coal.

For the analysis presented in this report, we use the long-run marginal emissions factor, which is based mostly on CCGT, to convert electricity taxes to carbon equivalent. In Appendix B, we present a version of the results using the grid-average electricity emissions factor instead.

Note that the issue of the conversion factor to use arises because many policies tax electricity downstream (on the use of electricity) rather than upstream (taxing the carbon content of fuels used to generate electricity). We return to this in the discussion of possible reforms in Chapter 8.

Because we are interested in price incentives at the margin, we want to estimate two things: first, how much the different policies increase the marginal price of each fuel for each end user (the price of consuming another kWh of gas or electricity, or another kg of coal or LPG); second, what the carbon content of a marginal unit of each fuel is. This is particularly important when thinking about electricity (see Chapter 6.3), which is generated from a range of fuels each with different carbon contents. During a certain load period the competitive electricity price is determined by the marginal production unit. In this study we assume that in the UK the marginal units are combined cycle gas turbines (CCGT) plants, consistent with other work (Redpoint, 2012). Thus, the carbon content of a marginal kWh of additional electricity consumption is assumed to be determined by the carbon content of gas-fired electricity.

There is, however, a debate as to whether, when thinking about the need for long-term, non-marginal reductions in consumption, carbon prices on electricity should be calculated according to the emissions factor of the marginal plant or according to the grid average emissions factor. This is considered in more detail in Box 6.1.

6.2 Policies included in the analysis

We include policies which have an effect on the marginal price of different fuels. These can be grouped in a number of different ways, as shown below.

Policies which raise energy prices directly

A number of policies are directly price-based, at either the energy production (upstream) or consumption (downstream) stage. Upstream pricing policies on energy inputs are the ETS and the CPSR. Downstream pricing policies on energy consumption are the CCL and the CRC. Large, energy-intensive industries that sign a CCA receive a discount on their CCL payments. We include the lower-rate of CCL in our analysis below but do not try to 'price' the carbon tax implicit in the industry-specific CCAs. Martin, de Preux and Wagner (2011) found the price penalty of CCAs to be modest.

We also include the Renewables Obligation (RO), which provides support for renewable energy by requiring providers to source a certain proportion of their energy from renewable sources. This raises energy costs at the margin. We use the buyout price and the overall size of the RO to calculate the implicit carbon tax, under the assumption that the buyout price is the effective marginal cost of the RO to energy providers.

Reduced VAT for households

Households pay a reduced rate of VAT on energy consumption (5% compared with the standard rate of 20%). This reduced rate applies to all energy types.

The reduced VAT rate amounts to an effective subsidy of 14.3% (1.2 ÷ 1.05) for households. However, to convert this to a carbon value, we need to know the price of domestic energy in order to establish the per-unit value of the subsidy and thus the implicit carbon price.

For electricity and gas, we use data from Department of Energy and Climate Change (2012h, i) for national average marginal prices in 2012, uprated to 2013 prices using out-turn RPI inflation rates for electricity and gas taken from the ONS. For coal, we use ONS estimates of the retail price of a 50 kg bag of coal as at April 2013. For LPG, there is no official statistic on which we can draw, and so we use the results of an Internet search for domestic LPG prices, taking the most commonly purchased size for domestic heating (a 47 kg refill container). Full details are given in Appendix B.

Policies delivered through energy companies

Our figures include three policies delivered through energy companies who recoup the costs from energy users. These are the Warm Home Discount (WHD), small-scale Feed-in Tariffs (FITs) and the Energy Company Obligation (ECO).

Whether these policies should be considered as 'carbon taxes' depends on how they are recouped (see Box 6.2). If they are recouped through fixed bill levies (such as higher standing charges), then they have no impact on marginal prices and so have no effective carbon price. On the other hand, if the policies are recouped through higher energy prices, then they act in the same way as other price-based policies and have marginal carbon incentives.[61] We follow Department of Energy and Climate Change (2013b) in assuming that the policies are recouped through marginal price increases, and so should be included in the carbon price calculations.[62] Table 6.1 summarises which policies are assumed to affect the marginal prices for different end-user and fuel combination.

Box 6.2: Policies funded through levies on energy bills

A number of the policies we consider (RO, ECO, FITs and WHD) are not taxes levied by the government, but are instead policies which deliver particular support for energy-related spending, usually administered by energy suppliers and paid for through levies on energy bills. As discussed above, if we assume that these policies lead to increases in marginal energy prices then we can think of them as equivalent to downstream taxes on energy use where the revenue is earmarked for particular forms of spending. For example, the RO provides a subsidy to producers of renewable energy, paid by energy companies and recovered through domestic and non-domestic electricity bills. We could imagine instead a policy which taxed electricity use directly at an equivalent rate to raise the same revenue, which is spent on subsidies for renewable generators. Similarly, the WHD provides a bill rebate to selected households, paid for by increases in household electricity and gas bills. We could tax gas and electricity and spend the money on rebates. Or we could fund them through carbon taxes applying to all fuels and users.

As described in Box 4.1, a number of these policies are already considered as effectively tax and spend measures in forecasting the public finances, and form part of the 'Levy Control Framework' agreed between DECC and HM Treasury. The exception is ECO: despite being funded through bill levies it is not part of the Framework.

Would it be preferable to switch from levy-funding to a system where the policies were funded through taxes on energy or carbon? At the very least, the mechanism by which policies are paid for would be clear. It is not at all transparent how energy suppliers recoup the costs. Although DECC assume that the full costs are covered through marginal price increases, there is little supporting evidence for this (House of Commons Energy and

[61] As we discuss in Chapter 2, though, there may be reasons to believe that even if the costs are recouped entirely through fixed bill levies, they would have some impact on energy use beyond a pure income effect and so could be treated as if they were increasing marginal prices.

[62] DECC estimate the price impact of these policies for households and for large intensive firms separate from other firms. DECC figures show no difference in the impact of these policies on energy prices for different firm types except for FITs, where the effect for non-energy-intensive businesses is given as £2/MWh and for intensive businesses as £0 to £2/MWh. To simplify the analysis we take £2/MWh for all firms. DECC also include other policies (smart meter roll out and 'better billing') as having marginal price effects in 2013. These effects are very small and we do not consider them here.

Climate Change Committee, 2013). The costs may be under- or over-shifted to consumers, and may be fully or partly recovered through a fixed charge on each bill rather than an increase in marginal prices. A fixed recovery is more akin to a poll tax on each household rather than a tax on each unit of energy (or carbon) consumed. A fixed levy would be more regressive than marginal price recovery, and would give no direct incentive to reduce energy use or emissions.

Further, funding the policies through a carbon tax would allow the costs to be met and provide consistent carbon incentives across fuels. At the moment, most of the policies are loaded onto electricity bills (and domestic bills). Where the beneficiaries are only domestic, it may make sense that the costs fall on the same group of energy users, but it is not really clear why the costs of coal or LPG should be unaffected.

Of course in general there is no particular reason why we would want to hypothecate revenues from a particular energy or carbon tax to be used to support renewables, microgeneration or pay for bill discounts for poorer households. Such earmarking (or 'hypothecation') has a number of disadvantages (Advani, Leicester and Levell, 2011). Most crucially, it is highly unlikely that the optimal spending on these measures is the same as the optimal additional revenue to be raised from energy or carbon. Setting the optimal mix of carbon or energy taxes, and spending the right amount on these other measures, are worthwhile in themselves but need not be tied together explicitly.

Table 6.1: Policies affecting different end-user and fuel combinations, 2013–4

	Households	**Small business**	**Medium business**	**Large energy-intensive business**
Electricity	ETS CPSR RO FITs WHD ECO	ETS CPSR RO CCL FITs	ETS CPSR RO CCL CRC FITs	ETS CPSR RO CCA rate of CCL FITs
Gas	WHD ECO	CCL	CCL CRC	CCA rate of CCL
Coal / LPG		CCL	CCL	ETS CCA rate of CCL

6.3 Results

Figure 6.2 shows the carbon prices implicit on electricity use for each end user. The price is broken down policy-by-policy with the combined total shown by the black dot. Policies in green are those which have direct price effects, those in purple are those delivered through energy companies, and the implicit VAT subsidy is shown in red.

Figure 6.2: Implicit carbon prices (£/tCO_2e) for electricity use, by end user (2013–4)

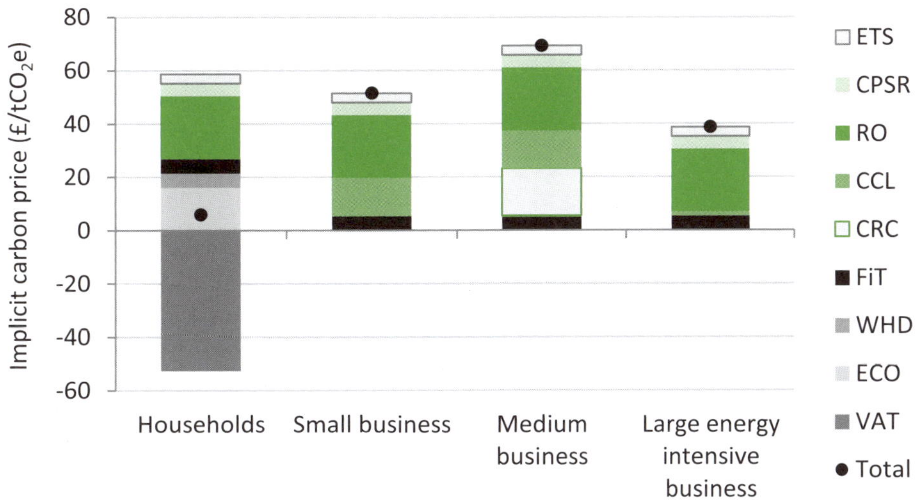

Source: Authors' calculations (see Appendix B for full sources and methodology).

Overall carbon prices for electricity use vary significantly across end users. Households pay £5.92/tCO_2e, large energy-intensive firms £38.71/t, small firms £51.53/t and medium-sized firms £69.17/t. Thus prices vary by a factor of around nine across different end users.

The policy-by-policy layering of effective carbon prices across end users is clear. There are a number of policies which are common across end users – the ETS, CPSR, RO and FITs. These collectively impose carbon taxes equivalent to £37.29/tCO_2e on all users (of which RO is the largest single policy, at £23.52/t).

It is variation in the other policies affecting different end users which gives the very different final carbon prices. Medium-size firms pay more than other businesses because they also pay the CRC, which adds £17.64/tCO_2e to their carbon price (see Table 5.2). Large energy-intensive firms benefit not only from not paying the CRC but also from a reduced rate of CCL, such that they pay a lower carbon tax even than small businesses.

For households, the biggest single effect is the reduced-rate VAT, which is an effective subsidy of £52.73/tCO_2e. This almost entirely outweighs the combination of other policies that raise domestic electricity prices at the margin. The latter includes a wider set of policies (WHD and ECO) which are recouped by energy companies through domestic energy prices but not business prices. Excluding the VAT subsidy, households would face a higher effective carbon price on electricity than small businesses or large energy-intensive firms (though still a lower rate than medium-sized firms).

Arguably another way of describing the taxation of energy at the household level is that the combination of other costs and taxes levied roughly outweighs the effect of the lower rate of VAT. It may have proved impossible up to now explicitly to levy VAT at the full rate, but a combination of much less salient policies has had much the same effect on bills. Of course given the need to reflect the cost of carbon this does not alter the case for, in addition, levying the full rate of VAT.

Figure 6.3: Implicit carbon prices (£/tCO_2e) for gas use, by end user (2013–4)

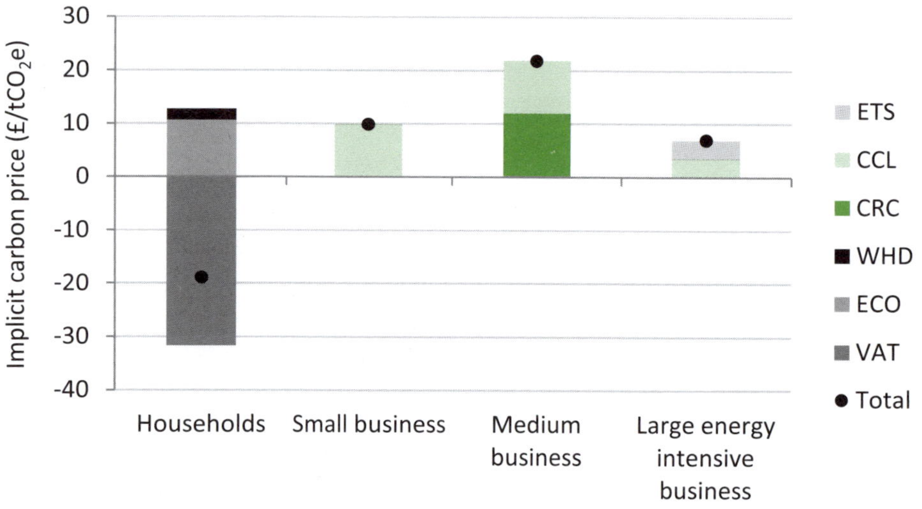

Source: Authors' calculations (see Appendix B for full sources and methodology).

Figure 6.3 repeats the analysis for gas use. What is notable is that there are far fewer policies which affect gas use than electricity use, such that overall carbon prices are lower for gas than electricity for each user. Despite this, the variation across end users is still present. Households pay a combined carbon tax equivalent to –£18.92/tCO_2e (i.e. an effective net subsidy), large energy-intensive firms £6.95/t, small firms £9.85/t and medium-sized firms £21.77/t.

For firms, only the CCL and CRC apply to gas use, and it is variation in the rate of CCL and eligibility for paying the CRC which drives the different prices faced. Households do not pay *any* direct carbon price on gas use, but do receive the effective subsidy from reduced-rate VAT (worth £31.68/tCO_2e on gas use) which is partly offset by the recouping of WHD and ECO from higher gas prices.[63] Again, excluding the VAT subsidy, households would face a higher implicit tax rate on

[63] Note that WHD and ECO are recouped through electricity *and* gas prices for domestic customers alone, whereas FITs are recouped only through electricity prices though for all end users.

gas use than small firms or large energy-intensive firms, but a lower-rate than medium-sized firms.

Figure 6.4: Implicit carbon prices (£/tCO_2e), by end-user and metered fuel type (2013–4)

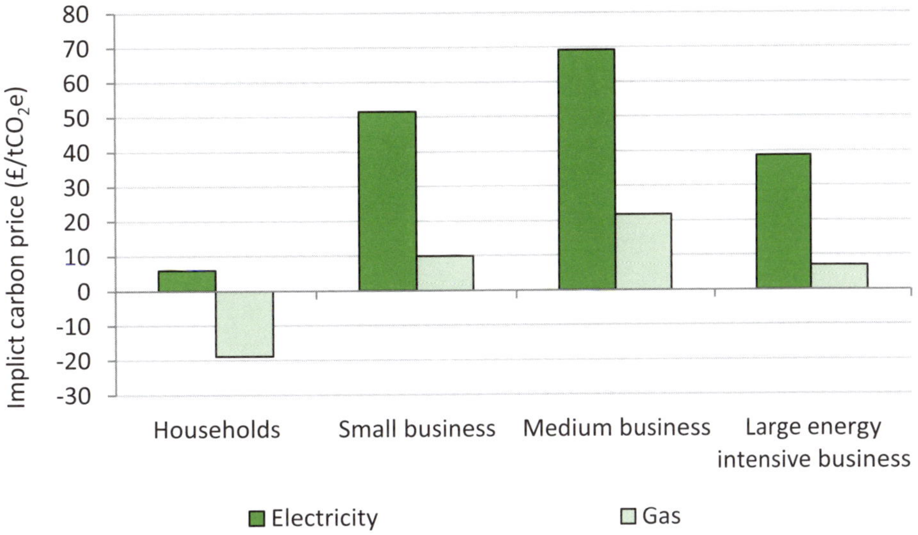

Source: Authors' calculations (see Appendix B for full sources and methodology).

Table 6.2: Implicit carbon taxes on electricity and gas use by policy, 2013–4

	Electricity (£/tCO_2e)	Gas (£/tCO_2e)	Notes
CCL	14.24	9.85	Tax levied on end-use of energy by firms. Difference is smaller based on grid-average electricity emissions but marginal incentives favour gas in carbon terms.
CCA rate of CCL	1.42	3.45	Larger discount for electricity (90%) than gas (65%) leads to higher implicit carbon tax on gas for firms signing a CCA.
CRC	17.64	11.92	Tax rate expressed as £12/tCO_2 but levied on energy *use* and converted on the basis of grid-average emissions. Leads to higher marginal tax rate for electricity.
WHD	5.34	2.13	DECC estimates 2013 price effect as £2/MWh for electricity and £0.40/MWh for gas. Households only.
ECO	16.83	10.64	DECC estimates 2013 price effect as £6/MWh for electricity and £2/MWh for gas. Households only.

Source: Authors' calculations (see Appendix B for full sources and methodology).

Figure 6.4 shows the overall electricity and gas prices for each end user to aid comparison across fuels as well as users. There is a considerable difference between gas and electricity within user group as well as the difference across user groups discussed above. The difference between implicit electricity and gas carbon prices amounts to £24.84/tCO_2e for households, £31.76/t for large energy-intensive firms, £41.68/t for small firms and £47.40/t for medium-sized firms.

This difference is partly about the policies which affect each fuel (the ETS, CPSR, RO and FITs have no impact on gas prices). It is also partly driven by variation in the size of the policy across fuels on a carbon basis. This is summarised in Table 6.2, which shows the marginal carbon-equivalent tax rate for electricity and gas for a number of policies where they apply to both fuels. In most cases (except the CCA rate of CCL), the variation favours carbon emitted through gas use relative to electricity use at the margin.

It is important to note that these figures show the implicit carbon prices associated with energy use policies. These should not be confused with the impact of these policies on energy bills. Box 6.3 contains a brief discussion of these bill impacts.

Box 6.3: The effect of energy use policies on prices and bills

The figures presented in this chapter show the implicit price associated with emitting a tonne of carbon *at the margin*. Intuitively, it is the amount saved for a small but sustained change in the amount of carbon emitted. However, from a welfare perspective one might be more interested in the effect that these policies have on average energy prices and bills.

Department of Energy and Climate Change (2013b) estimates that all energy and climate change policies add £2/MWh (5%) to average household gas prices and £22/MWh (17%) to average household electricity prices in 2013 (all price in 2012 values). This is set to rise to £3/MWh (5%) for gas and £49/MWh (33%) in 2020.

The cost of these policies accounts for £112 (or 9%) of an average household dual-fuel energy bill in 2013. However, this does not take into account the expected benefits in the form of greater domestic energy efficiency (and therefore reduced energy consumption) that potentially occur as a result of these policies. When these benefits are included, DECC estimates that the policies result in average bill reductions (relative to a baseline where no policies are in place) of £15 (1%) in 2013 and £84 (6%) in 2020. We have not attempted to assess the accuracy of this claim.

These policies are estimated to increase average business energy bills by between 2% and 21% in 2013, and 10% to 30% in 2020.

6.4 Future carbon prices

The results so far have considered current (2013–4) carbon prices across fuels and users. A natural question is whether, on current policies and forecasts, the variation in prices is set to widen or narrow in the future. We estimate how carbon prices will look in 2020 using a range of sources for how policies and prices are expected to evolve (see Appendix B for details). If anything, the range of prices is set to widen rather than narrow. All figures are presented in 2013 prices unless otherwise stated.

Table 6.3 shows the relative electricity and gas carbon price across end users (i.e. the ratio between them, such that 2 suggests electricity carbon prices are double those for gas) as well as the absolute difference in terms of £/tCO_2e. Prices for 2020 in the table and throughout this discussion are expressed in 2013 terms.

Currently the electricity carbon price is between three and six times the gas carbon price, depending on user.[64] By 2020 this will rise to five to eleven. In cash terms the cost of emitting a tonne of carbon dioxide is £25 to £47 more expensive from electricity than from gas. This will rise to £55 to £101 (in 2013 prices) by 2020.

Table 6.3: Relative carbon prices between gas and electricity over time, by end user

	Year	Households	Small business	Medium business	Large energy-intensive business
Ratio (electricity:gas)	2013	N/A	5	3	6
	2020	N/A	11	5	8
Difference, £/t (electricity – gas)	2013	24.84	41.68	47.40	31.76
	2020	55.36	100.67	95.50	82.13

Source: Authors' calculations (see Appendix B for full sources and methodology). No ratio figures are given for households since the carbon price for gas for households is negative. Ratio figures are rounded to the nearest whole number. 2020 figures in 2013 prices.

Table 6.4 shows current and 2020 carbon prices for electricity and gas by end user (all in 2013 prices). Future policies will see large increases in carbon prices from electricity use but have relatively little impact on gas, leading to the increasing divergence in carbon prices over time across fuels. The pattern across users is similar in the two years, with the highest prices faced by medium-sized business – those who are covered by the CRC – whilst households continue to face much lower prices.

[64] This ratio is not well defined for households, since they face a negative carbon price for gas.

Table 6.4: Implicit carbon prices (£/tCO$_2$e) by end-user and fuel type, 2013 and 2020

	Year	Households	Small business	Medium business	Large energy-intensive business
Electricity	2013	5.92	51.53	69.17	38.71
	2020	26.64	110.51	121.24	94.40
Gas	2013	−18.92	9.85	21.77	6.95
	2020	−28.72	9.85	25.74	12.27

Source: Authors' calculations (see Appendix B for full sources and methodology).

The increase for electricity comes largely from the rise in CPSR values and the introduction of Contract for Difference Feed-in Tariffs (CfD FITs), the latter replacing the RO as the main support for renewable generation. CfD FITs (together with remaining RO costs) are expected to equate to around £45 to £50 per tonne of CO_2 for all end users in 2020,[65] compared with current RO costs of around £23.50/tCO$_2$e. The CPSR in 2020 is equivalent to around £21/tCO$_2$e compared to current values of just under £5/tCO$_2$. There is some offsetting effect from the CRC, which falls in cost from around £18/tCO$_2$e to £11/tCO$_2$e.

Figure 6.5: Implicit carbon prices (£/tCO$_2$e) by end-user and fuel type, 2013 and 2020

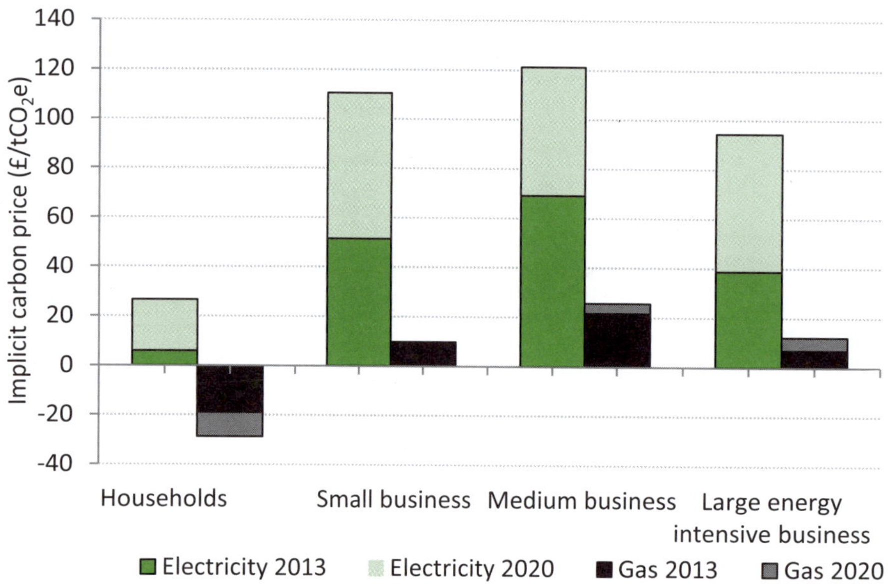

Source: Authors' calculations (see Appendix B for full sources and methodology). Figures in 2013 prices.

[65] Note that our estimates are based on Department of Energy and Climate Change (2013b) estimates of the price impact in 2020, which are slightly higher for households than for firms leading to some variation across end users.

For households much of this is offset by the rising value of the VAT subsidy as the unit price of electricity rises. This increases the variation across domestic and non-domestic carbon prices. In 2013, the electricity carbon price faced by households is between £28 and £65/tCO_2e, lower than that faced by firms, depending on firm size. The gas carbon price is £25 to £40/t lower. By 2020 the difference rises to between £67 and £95/t for electricity and between £38 and £55/t for gas.

The carbon price for electricity is also affected by the decarbonisation of the electricity grid. By 2020, the carbon content of a marginal unit of electricity is projected to have fallen by a fifth from 0.368 tCO_2e /MWh in 2013 to 0.293 tCO_2e /MWh[66]. The content of the average unit is expected to fall even more dramatically from 0.494 tCO_2e/MWh in 2013 to 0.196 tCO_2e/MWh in 2020. Policies which charge a constant price per unit of electricity consumed, such as the CCL, will therefore be imposing a higher price per unit of carbon, but on a shrinking base. Policies which charge a constant price per unit of carbon, and then use a conversion factor based on average emissions, such as the CRC, will be charging a lower price per unit of carbon, as the carbon content of the average unit falls more quickly than that of the marginal.

For gas the only increase in carbon price comes from an increase in the CRC price that is due to be implemented next year. After this there are no currently announced policies that will increase the carbon price in real terms, leading it to become relatively cheaper than electricity in carbon terms. For households, expected increases in gas prices will increase the real value of the implicit subsidy from reduced-rate VAT. Future changes to gas prices are however highly uncertain and difficult to predict. This means that the change in the size of the subsidy is unclear, and so the impact on future carbon prices is also uncertain.

It is important to note that without the implicit VAT subsidy, current policy would result in an effective carbon price that is approximately equivalent to the targeted non-traded carbon price of £59 /tCO_2e in 2013. As a result, removing the reduced rate of VAT on domestic energy and replacing it with the full rate of 20% would achieve this carbon price. We illustrate such a reform in Chapter 8, where we discuss in detail the effects this would have on carbon prices and the ways in which potential distributional consequences could be mitigated.

This will no longer be the case in 2020 due to the growing effect of a number of policies on the effective carbon price. If the VAT subsidy were removed, the projected carbon price for 2020 would far exceed the target price of £65/tCO_2e under current policy plans. This is true across households and firms of all sizes.

It is perhaps surprising that current policy plans lead to carbon prices that exceed the stated non-traded carbon price by such a considerable amount in 2020. This raises the possibility that policy is in practice aimed at realising a (much) higher effective carbon price. For example, the 2030 decarbonisation

[66] This is both because gas generation is forecast to become more carbon efficient and because renewable generators will sometimes be the marginal plant.

target may require greater savings than implied by the target price of £70/tCO_2e in 2030, and so require a trajectory of higher carbon prices. If this is indeed the case, and target carbon prices are set at a higher level, policy would require fewer future changes in order to reduce future carbon prices to this level.

6.5 What is the right carbon price for the UK?

The above analysis has shown that the carbon prices facing energy users in the UK are inefficient. Notably, there are wide disparities in the carbon prices of electricity and gas, and between the prices facing households and businesses. These differences are set to increase in the future. In developing the case for reform in the next section, we need to consider one further, fundamental question – what is the right carbon price to target through policies in the UK?

As noted in Chapter 3, the government uses a target-consistent approach to answering this question, under which the value of carbon is set at the level required to meet prevailing emissions reduction targets. This has led to two sets of carbon values: one (the traded price of carbon) applying to emissions occurring in sectors covered by the EU ETS, such as emissions from electricity; and the other (the non-traded price of carbon) occurring in sectors not covered by the EU ETS, such as gas. The box below considers why government distinguished between the two values and how they were calculated.

Figure 6.6: A comparison of the implicit carbon prices for gas (£/tCO_2e) and the non-traded sector carbon price of £59/ tCO_2e, by end user

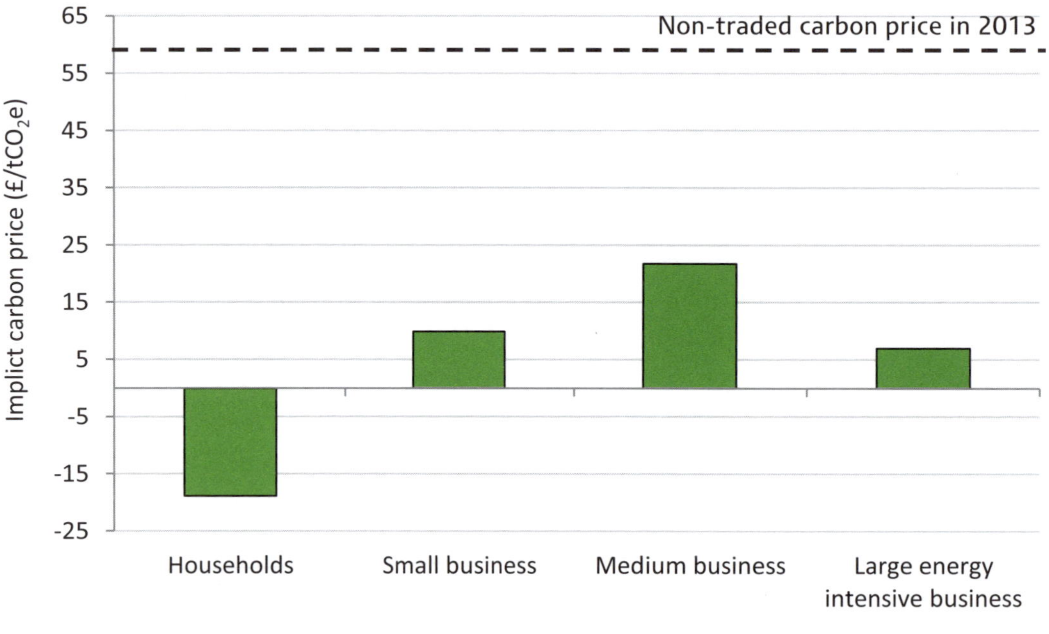

Source: Authors' calculations (see Appendix B for full sources and methodology).

In 2013 the target price for carbon in the non-traded sector is £59/tCO_2e, rising to £65 in 2020 and £76 in 2030. This provides an appropriate set of values for assessing the adequacy of the carbon prices imposed by current policies on gas and other uncapped fuels such as coal and LPG.

As shown in Figure 6.6, gas prices paid by all users are well below the efficient level consistent with meeting emissions reductions targets in the non-traded sector. The price paid by households is roughly £80/tCO_2e below this figure, while even the price paid by medium-sized businesses - who by some way face the highest carbon prices for gas - is roughly £40/tCO_2e below the efficient level. We draw on this analysis in making recommendations for reform in the Chapters 7 and 8.

Box 6.4: Why distinguish between the traded and non-traded price of carbon?

The reason for distinguishing the traded from other sectors of the economy is that in practice they are subject to two different emission reductions targets.

Within the traded sector, emissions are capped by the EU ETS, such that, so long as the ETS cap is binding, any reduction in UK emissions in the traded sector will not reduce global emissions, but displace those emissions elsewhere in the EU. The traded price of carbon is driven by the ETS cap and is equal in the short term to the EUA price.

In the non-traded sector, in contrast, the emissions reduction target is that imposed by UK carbon budgets. The non-traded price is calculated based on estimates of the marginal cost of reducing emissions to comply with the budgets. Because UK carbon budgets currently effectively impose a more stringent constraint than the ETS cap, the non-traded price of carbon as calculated by government is much higher than the traded price.

From 2030 onwards, emissions reduction targets are less well specified and Government has set a single set of carbon values, based on global abatement costs consistent with a central expectation of stabilising temperature increases at no more than 2 °C above pre-industrial levels. This implicitly assumes a comprehensive international agreement to address climate change based on a global carbon market.

In 2013 the traded price is only £6/tCO_2, rising to only £9/tCO_2 in 2020 before increasing rapidly to £76/tCO_2 in 2030. The very low values up to 2020 reflect the recent collapse in the value of EUAs, while the rapid increase to 2030 reflects the fact that much more stringent (global) emissions reductions will be required in the 2020s in order to be on track to keep expected temperature increases within manageable levels.

Using these values as a way of identifying the appropriate carbon price for electricity is more problematic than was the case for non-traded sector fuels. If we were only interested in short-term costs and benefits, we might be tempted to conclude that UK electricity users should be facing a carbon price of electricity of only £6/tCO_2e in 2013 – i.e. that there should be no separate UK electricity carbon taxes today and a sharp increase in carbon prices in the 2020s.

However, such a trajectory – very low carbon prices today and a very significant increase in the 2020s – is unlikely to represent a least cost path to the levels of electricity decarbonisation required by 2030 to address climate change (Committee on Climate Change, 2013c). Indeed, this was an explicit rationale for the UK government introducing the Carbon Price Floor, which rises steadily to £70 in 2030.

It is, therefore, more difficult to draw clear conclusions about the appropriate carbon price for electricity in the short term. Certainly there is no compelling case for carbon prices higher for electricity than for gas that are currently seen in the UK. But arguably a smoother trajectory to 2030, like that implied by the steady increase in the non-traded price of carbon, would be more appropriate than the very low short-term prices followed by sharp increases given by the traded price of carbon. This perspective is reflected in our recommendations in Chapters 7 and 8, which would see the carbon prices of electricity and gas equalised at around £59/tCO_2e in 2013.

It is important to recognise, however, that as long as the implicit carbon price for electricity in the UK is higher than the price of EUAs, this will involve higher cost abatement in the UK, with no aggregate reduction of emissions at an EU level. It will also serve to put further downward pressure on the price of EUAs. That is why domestic pricing of electricity must be accompanied by efforts to negotiate a more stringent ETS cap and put in place mechanisms to tighten the cap should the price fall below a certain level. Again, we consider this in the following chapter.

6.6 Summary and conclusions

The use of carbon prices by policymakers recognises that increasing the cost of carbon emissions will encourage end users to reduce their emissions. However, the incentives provided by the current set of policies do not encourage this reduction to happen in the most efficient way. Even restricting our attention to the two main metered fuels, electricity and gas, we find substantial variation in prices across users and, for a given type of user, across fuels. This has come about through the gradual layering of different policies that impose implicit carbon prices but affect different fuels and users. If anything, the variation is set to get quite a lot larger in the next few years, as additional policies, which affect electricity prices but not gas prices, are introduced. If we include non-metered fuels (see Appendix B) then the range of prices simply gets even larger.

Variation in carbon prices across users at least partly reflects a conscious choice by policymakers to try to protect domestic energy use from policies which directly raise marginal prices, though a number of (perhaps less transparent) policies which are funded through levies also act as effective carbon prices and largely affect the domestic sector. Including them reduces the variation across end users in carbon prices, but there is still some doubt as to whether or not they have the marginal price effects which are assumed. In any case, the position for households is overwhelmed by the very large subsidy they receive from paying a reduced rate of VAT on energy use. As energy prices are forecast to rise in real terms in the future, the value of this subsidy will go on rising.

Variation also reflects perhaps less conscious choices to layer policies much more heavily on electricity use than any other fuels. In the future, as the expected marginal carbon content of electricity falls, the gap between electricity and other fuels will rise. This makes clear the rationale for imposing carbon prices upstream on the inputs to electricity generation, rather than on the electricity consumed. In the next chapter we propose a set of reforms that would ameliorate some of these issues in the shorter term. We also set out how a longer-term reform of policy might look.

7. Recommendations for upstream and firm energy use policies

Samuela Bassi, Alex Bowen and Sam Fankhauser

In Chapter 1 we noted that, where possible, policy should attempt to achieve a uniform carbon price across users and fuels to provide incentives for efficient investment in emissions abatement. In addition, this would provide incentives for efficient investment in emissions abatement. It should also help with the provision of a stable and credible path for future carbon prices. Where there are other reasons for intervention, such as equity concerns or worries about carbon leakage, it is in general better to tackle these with separate policies focused on the particular problem, rather than varying carbon prices for different groups. Additional objectives, such as decarbonising electricity generation, may require further targeted support beyond that which would come from efficient pricing of carbon. A further objective is to keep administrative and reporting costs low, without jeopardising compliance.

In the light of the economic principles and the policy objectives set out in Chapter 1 and 2, the discussion in Chapter 4 highlighted aspects of the current policy landscape. We then set out in detail in Chapter 6 what current policies imply for the variation in effective carbon prices across fuels and end users, and how this has changed in recent years and may change in the future given current policy trajectories and fuel price expectations.

Taking all these factors on board, this section considers options for policy reform. We begin (in Chapter 7.1) with policies which affect electricity prices 'upstream', at the point of generation, but the effect of which is passed through to users. We then look at policies imposed separately on energy used by firms (Chapter 7.2) and households (Chapter 7.3) 'downstream', at the point of use. This division, which already exists in the current policy environment, allows for some separation between efficient carbon reduction objectives, which should best be met through upstream prices common across users, and other issues, such as energy efficiency and fuel poverty, which can be tackled by more targeted policies. We consider both shorter term practical changes that could be made within the current policy environment, and longer term reforms.

7.1 Reforms to upstream carbon prices

The key policies which affect the upstream carbon price, and hence affect all users of electricity, are the EU ETS, the Carbon Price Floor (CPF) and the Renewable Obligation (RO) – the latter soon to be replaced by the Feed-In-Tariffs with Contract for Difference (CfD FITs). In this sub-section we present a number of policy recommendations focusing on the impact of these policies on energy users, rather than on the functioning of the electricity market *per se*.

The EU ETS imposes a price based on the carbon content of fuels used in the generation of electricity and by other carbon-intensive industries, and so provides a common price across users and fuels. However, the price of carbon in the EU ETS has fallen to a level that could damage not just Europe's low carbon investments, but also the credibility of the ETS itself. The problem is largely structural and inherent in the way the ETS has been designed, which has led to over-allocation of allowances. Back-loading if implemented, may delay the problem of oversupply, but is unlikely to resolve it unless the surplus permits are eventually retired. Renewables policies, such as the RO, can also have undesirable spillover effects on the EU ETS. By reducing domestic emissions from traded sectors, they can depress the price of EUAs rather than reduce the overall amount of emissions covered by the scheme, which is fixed by the cap (Fankhauser, Hepburn and Park, 2010).

In the medium to long term more radical structural reforms are needed to improve the functioning of the ETS. As the Phase IV negotiations for the ETS are under way, the UK government should negotiate strongly for a much more stringent cap. The government should also support the introduction of a floor price mechanism into the overall ETS. By permanently reducing supply through the removal of permits, a floor price would cause the EUA price to rise. A 'ceiling' could also be introduced, so that if the price rises above some higher threshold, any set-aside permits may be reintroduced to help control the cost of emissions reduction. Several mechanisms to introduce price containments are possible, with different implications in terms of regulatory requirements (see, for example, Taschini, 2013). Together, the floor and ceiling could evolve in line with emissions caps in future phases so as to set a credible path for future ETS carbon prices that should increase over time.

Additionally, it is recommended that, from Phase IV, as far as possible all permits should be auctioned, so that the rents from the policy go to government rather than being captured by firms. However, this is contingent on having some alternative policy to deal with the issue of leakage.

The Carbon Price Floor (CPF) further supports the EU ETS price signal in the UK. This provides a trajectory for the carbon price by charging the difference between the future allowance price (in two years) and the pre-announced CPF (this difference is the CPSR).

Should future European policy reforms be introduced, the CPF could effectively be superseded by EU legislation. In the short term, there is value in retaining a unilateral support to the carbon price in the UK (while recognising this is very much second best to an effective pricing system at the EU level). However, the way the CPSR is determined could be reviewed to ensure its carbon price signal, in combination with the ETS price, is closer to the CPF target.

To the extent that the trajectory of the CPF is credible, it should provide sufficient certainty to energy users considering investment decisions while the future phases of the EU ETS are still under negotiation. However, given the experience with the fuel duty escalator, there are reasons why one might not expect that the CPF trajectory set out for future years will be delivered. By introducing the proposed CPF rate into primary legislation, a government would, to some extent, be able tie the hand of its successors by at least requiring a parliamentary vote in order to make changes. This might be seen as a stronger constraint than allowing the rate to simply be confirmed or adjusted by the Chancellor at future Budget statements.

7.2 Reforms to policies which affect energy used by firms

Currently energy use by firms is affected 'downstream' by the CCL, CCA and the CRC, as well as by indirect policy costs passed on by energy suppliers through electricity prices, such as the costs of the EU ETS and the CPF. This chapter focuses on the reform of downstream policies, with the aim of unifying carbon prices across firms and fuels. A key constraint in doing this will be finding a solution to the issue of carbon leakage, as we discuss later in this Chapter.

Simplifying the policy landscape

The layering of policies affecting energy use by firms has led to significant policy overlap, so firms pay for carbon through a number of policies. The interaction of different policy instruments across sectors has led effective carbon prices to vary significantly across the economy (as illustrated in Figure 6.5).[67] This leads to inefficient allocation of abatement activity across sectors, distorting relative prices of final goods and services.

Layering policies also creates institutional complexity, adding to the administrative burden for firms. In part this is by design. The CRC, for example, aimed not only to increase the carbon price for firms, but to improve its salience. However, it is not clear that this effect is worth the additional inefficiency caused by increased administrative costs. This suggests that emissions could be reduced more cost-effectively by simplifying the existing policy landscape. Ultimately, all downstream carbon and energy taxes could be reduced to a single instrument applying an even carbon price across all sectors, though there are also a number of reforms which could be made to the current set of policies.

First, the history of frequent revisions to the mechanisms, monitoring and eligibility criteria of the CRC mean that there is significant uncertainty about the future of the policy, weakening the abatement incentives provided. Furthermore, the price signals provided by the CRC, if they are wanted, could easily be combined into the CCL, reducing much of the administrative burden.

Furthermore, the CCA was introduced to reduce potential leakage as a result of the higher energy costs imposed by full CCL rates. However, the current eligibility

[67] Figure 6.5 slightly understates the variation as it omits the small number of firms who qualify for CCAs but are covered by the CRC. A very small number of firms are also exempt entirely from the CCL.

criteria are rather broad, and include sectors (such as retail, for example), for which the potential for carbon leakage appears small. *Ex post* studies based on firm level data have found effectively no impact of the full CCL on the output and employment of firms, compared with the performance of firms under the CCA regime (Martin, de Preux and Wagner, 2011). This suggests that the CCA may not be justified on the grounds of competitiveness alone. The CCA is also associated with expenditures (in terms of time and monetary costs) arising from the negotiations between the public administration, individual industry bodies and firms over the terms of the CCAs themselves.

In contrast, the CCL has proven relatively effective in reducing energy intensity in medium firms, while no evidence was found of substantial impact on employment, gross output or productivity. Taken together, the existing evidence would point towards replacing the CRC and the CCA with a single, revised CCL. *Ex ante* theoretical analysis, however, suggests that, under certain assumptions, leakage risks from raising carbon prices (as would be the case if we move from the CCA to a full CCL) may still exist (see, for example Monjon and Quirion, 2009; Ritz, 2009; Varma et al., 2012; Grubb and Counsell, 2009). It will be crucial therefore that any policy reform is assessed against these risks in more detail. At the very least, efforts should be made to better identify those firms most vulnerable to leakage and the eligibility criteria adjusted to reflect this.

Carbon price unification across fuels and firms

The long-term aspiration for policy should be a uniform carbon price for all sectors and fuels that is levied upstream. However, this is unlikely to be feasible in the short term. This section focuses on how downstream policy reforms could move towards a uniform carbon price.

A revised 'CCL+', which replaces the CCL, CCAs and CRC, might look as follows:

- a CCL+ would be levied on all businesses, including small activities that are currently exempted from the CCL. Tax rates should be revised so that the same implicit carbon price applies to all main fossil fuels, that is, natural gas, coal and LPG.
- The CCL+ rate for electricity should be lower than for other fossil fuels, since electricity is already subject to an implicit carbon price associated to upstream policies (notably the EU ETS, CPF, RO/CfD and FiT), whose cost are passed on to electricity users by energy suppliers. This should be calculated as the residual between the CCL+ carbon price charged on gas, coal and LPG minus the implicit carbon rate of upstream policies.
- For sectors covered by the EU ETS, the carbon rate on natural gas, coal and LPG should be reduced by the ETS allowance price, which firms pay under the trading scheme. In order to determine what the CCL+ tax rate should be, a mechanism conceptually (although not operationally) similar to the current CPF could be adopted (see also recommendations in Chapter 7.1).

The current and proposed policy regimes are outlined in Table 7.1.

Table 7.1 Overview of current policy regime and proposed policy reform for the power, traded and non-traded sectors

Sector	Policy application	Current policy regime		Proposed policy reform	
		Electricity	*Other fuels*	*Electricity*	*Other fuels*
ETS industries	Direct	CCL/CCA	CCL/CCA, ETS	CCL+	ETS, CCL+ (minus ETS)
	Pass-through	*ETS, CPSR, RO, FIT*	—	*ETS, CPSR, RO, FIT*	—
Non-ETS sectors	Direct	CCL/CCA, CRC	CCL/CCA,CRC	CCL+	CCL+
	Pass-though	*ETS, CPSR, RO, FIT*	—	*ETS, CPSR, RO, FIT*	—

Setting the right CCL+ rate will be a sensitive issue. In principle, there is an argument for aligning the carbon price of the traded and non-traded sectors along a single value. This should ensure that GHG externalities are priced consistently across all fuels and sectors, at a level consistent with the UK carbon budgets. The Government envisages that this will occur in 2030, when the carbon price is expected to reach £76/tCO_2e.

One way to reach this target would be to gradually align the CCL+ to the government's estimated carbon price trajectory for the non-traded sector, which is consistent with the UK carbon budgets. As noted previously, this is £59/tCO_2e in 2013, rising to £65 in 2020 and £76 in 2030[68]. This, however, may be a considerable increase for ETS traded sectors and energy-intensive firms benefiting from reduced energy tax rate through the CCA. Other intermediate targets could be considered, in order to smooth the transition from the current relatively low carbon rates. The Government should devise a clear and coherent trajectory for the carbon price, in consultation with the business sectors and interested stakeholders.

Figures 7.1a and 7.1b display how an illustrative policy reform would apply across different fuels and sectors should CCL+ rates aligned to the non-traded sector price be applied in 2013 (for simplicity only gas and electricity are included). The proposed policy regime would result in a substantial tax increase on gas and other fuels, as the current carbon prices for these are significantly lower than the target carbon price for the non-traded sector. For electricity, the increase will be lower, and for some firms taxes will fall. This change will depend on the combination of CCL, CCA and CRC currently faced by the firm. For instance, under the CCL+, installations currently subject to both the CCL and the CRC (medium businesses) would see a small reduction in their electricity tax rate.

[68] https://www.gov.uk/government/uploads/system/uploads/attachment_data/file/68944/Tables_1-20.xlsx

This would fall by around £7/tCO_2e, or 2% of the current electricity bill, compared with the present policy regime. Installations subject to the CCA (large energy-intensive businesses) would see an increase of around £22/tCO_2e, around 10% of the current electricity bill.

Figure 7.1a Illustrative reform of electricity taxation in 2013

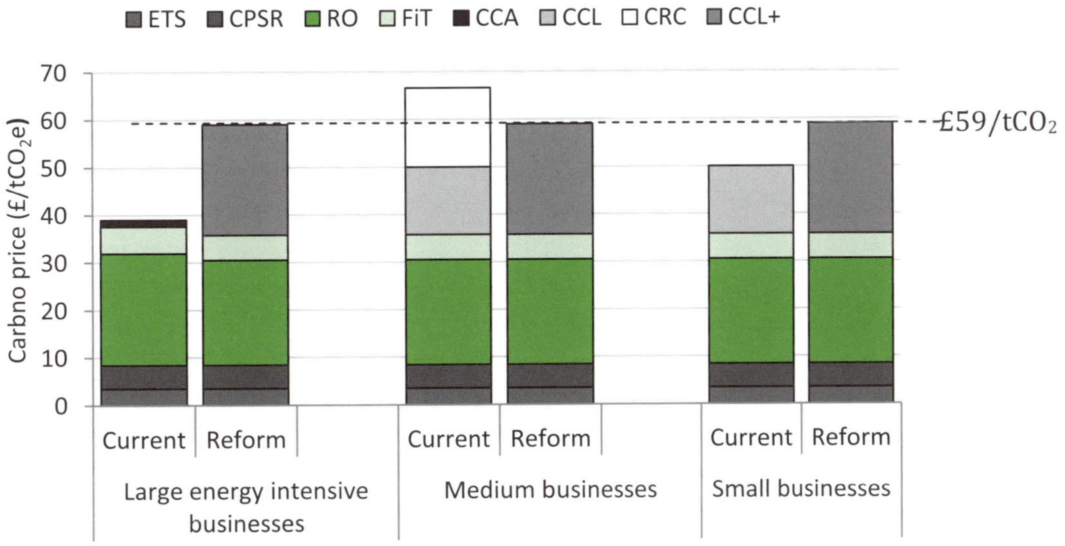

Figure 7.1b Illustrative reform of gas taxation in 2013

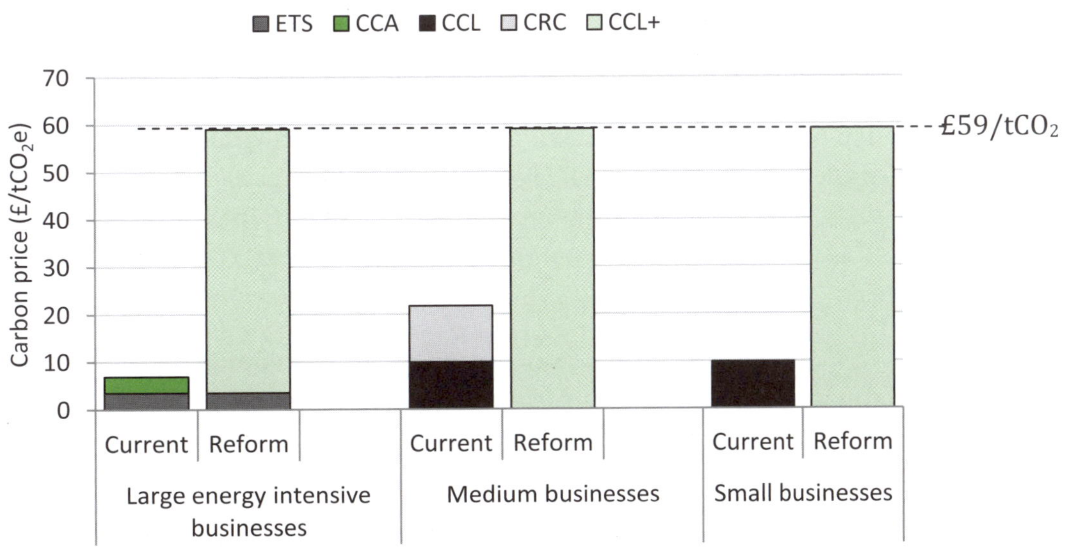

Source: Authors' calculations. For sources of current tax rates and conversion factors see Appendix B. Note: Large energy-intensive businesses are assumed to be subject both to the EU ETS and to be part of a sector which is eligible for a CCA. In practice, there are some small firms that are directly affected by the ETS, but do not have a CCA. For these firms, the CCL+ rate would be adjusted to ensure that the implicit carbon price does not exceed the target price. Medium businesses are assumed to be subject to both the CRC and the full CCL rates. Small companies are subject to the CCL only. Very small companies which are not subject to the CCL are not included in the charts.

As the policy costs passed on through electricity prices (ETS, CPF, RO/CfD and FITs) increase over time, the CCL+ rate on electricity should be adjusted periodically so that the sum of the cost pass-through and the CCL+ rate for electricity matches the carbon price target for the year. By 2020, the cost pass through from upstream policies alone would have overtaken the target non-traded carbon price for that year (£65/tCO_2e). In such a case the CCL+ rate for electricity should be zero. Compliance with the current EU Energy Tax Directive (2003/96/EC), however, may require electricity to be taxed at least by 0.5 €/MWh for use in the business sector.

A careful examination of the impact of upstream policies on the carbon price would be required to ensure that carbon prices do not greatly exceed the target non-traded price. Where policies have alternative objectives to pricing carbon (such as R&D and network externalities in the case of CfD), it may be worth considering alternative options for funding such policies in order to mitigate the effects of these policies on the effective carbon price.

In the longer term, we suggest replacing the CCL+ with a single carbon pricing policy imposing a uniform carbon taxes on coal, gas and LPG further upstream - i.e. at the point of import or manufacture. Levying upstream taxes on fuels can reduce administrative costs and scope for tax evasion. Whether a downstream taxation can be entirely replaced by such an upstream instrument will also depend however on the requirements set by the European Union legislation that set taxes at the point of business use (such as the Energy Tax Directive (2003/96/EC) and its future revisions).

The issue of carbon leakage

The illustrative policy reform described above shows how aligning carbon prices to the non-traded sector values will result in some companies paying higher energy tax rates than they currently face, while others may pay less. The most affected will be energy-intensive companies, who currently face the lowest carbon prices.

The change in tax regime would also result, on balance, in a net increase in fiscal revenues compared to the revenues raised by current policies. Further details on the potential revenues raised by such reforms (should no compensation mechanism be introduced), and the effects on emissions, are discussed in our companion paper (Bassi, Dechezleprêtre and Fankhauser, 2013).

Overall, we recommend that a policy reform along the lines here suggested should be kept revenue neutral, and the extra revenues recycled to mitigate the impact of the policy change on the sectors most hit by it.

A key concern that revenue recycling should address is the risk of carbon leakage. The reforms we propose have to account for the fact that, at least initially, Britain's trading partners may have weaker carbon pricing policies. This means that over the medium term energy-intensive industries that are trade-exposed may require additional protection.

The extent to which carbon prices can lead to, or have already induced, carbon leakage is still unclear. A look at the literature reveals that '*ex ante*' theoretical studies generally point to fairly substantial rates of carbon leakage. In contrast, '*ex post*' empirical analyses generally fail to find convincing evidence of substantial leakage.

No detailed investigation of carbon leakage has been attempted for the policy reform illustrated above, and further detailed analysis will be needed to identify which sectors will be most at risk. In some cases the array of sectors that have been eligible to some form of mitigation of carbon prices appears to have been too broad. This should be avoided in future. For example, the CCA is available to firms which are either energy intensive *or* highly trade exposed. This has included sectors which are hardly at risk of carbon leakage, such as large retail or baking.

Evidence also suggests that the ETS criteria used to identify sectors at significant risk of carbon leakage, which in turn is used to assign auction exemptions, have been overly generous (see, for example, Martin, Muûls and Wagner, 2012). Under the current regime, sectors that are highly trade intensive qualify for exemptions regardless of their carbon intensity. In fact, Martin, Muûls and Wagner (2012) show that trade intensity is the route by which most sectors qualify, and argue that the aggregate risk of carbon leakage resulting from an application of the current EU ETS exemption criteria could be achieved with just a fraction of the amount of permits that will be handed out for free. While the identification of sectors effectively at risk of carbon leakage falls outside the scope of this study, we recommend that the government should investigate more accurate approaches for sectors eligibility. It will be crucial to revise the normative framework for industry compensation to ensure that the right sectors are targeted.

Furthermore, adequate compensation mechanisms, both for the EU ETS for traded sectors and for a CCL-style energy tax for others, will have to be carefully investigated. Clearly, the 'first best' option would be to achieve an agreed global carbon price, which would remove any potential issue of leakage associated with carbon pricing. A 'second best' alternative would be to implement some form of border adjustment (e.g. a charge on imports), so that domestic and imported goods would be subject to the same carbon price. None of these options, however, is likely to be feasible in the short term. A range of 'third best' options may therefore be required. These could include free allowances under the EU ETS and revenue recycling from other carbon pricing measures (such as the proposed CCL+) as long as this does not create variation in the carbon price across sectors.

Once the sectors most at risk of carbon leakage are compensated, it is possible (depending on the tax rate ultimately chosen by the government and the response of firms to such a policy) that some revenue may remain from the revised carbon pricing policy. In this case, this could be recycled back as direct compensation to all sectors, for example by reducing other taxes.

8. Recommendations for household energy use policy

Arun Advani, Paul Johnson, Andrew Leicester and George Stoye

The previous chapter considered potential reforms to policies which affect energy use by businesses. Difficulties in accurately identifying the extent to which leakage is a genuine problem and identifying which sectors are most at risk makes it extremely difficult to model very specific recommendations.

This is not an issue for the household sector and we can model more specific reforms. In this chapter, we consider reforms around policies towards domestic energy use in four broad areas – pricing, support for renewable generation, support for energy efficiency and support for energy bills.

8.1 Energy pricing

Households pay VAT on domestic energy at a reduced rate of 5% rather than the full rate of 20%. This implicit subsidy of more than 14% to energy prices costs in excess of £5 billion per year. While we recognise the political constraints here, and understand that politicians will remember the effects of the last attempt to impose VAT at the full rate back in 1993, it is important to recognise that this effective tax rebate has the potential to undermine the effectiveness of carbon pricing and other energy efficiency policies.

Whilst the VAT subsidy applies to all fuels, policy towards domestic consumption has been inconsistent across different forms of energy. Households do face implicit carbon price signals for electricity use, both because of the upstream policies that feed into consumer prices and because of a number of policies which have costs that are recouped through electricity bills. In carbon terms, households face prices that are almost £25 per tonne higher for electricity than gas, and the gaps are even larger for non-metered and highly carbon-intensive fuels such as coal.

There is therefore a clear case in the medium term to move towards more consistent carbon prices for the domestic sector, both across different fuels used by households and compared with carbon prices levied on business energy use. This could be achieved through imposing VAT on domestic energy use at the standard rate, and levying new taxes on gas and non-metered fuels that broadly align the implicit carbon prices faced by households for consuming different fuels.

Of course such reforms would be significant. Aside from the £5 billion or so raised from extending VAT to 20% on domestic energy, a new gas tax of 0.8p/kWh (roughly the level needed to equate implicit gas and electricity carbon taxes at the moment) could also raise around £3.3 billion[69]. Together, they would raise electricity prices by 14% and gas prices by 34%, as shown in Table 9.1 (replicated from Advani et al., 2013). The fact that such increases are far from unprecedented (electricity prices rose by 15% between August 2011 and May 2013, gas prices by 33% between November 2010 and May 2013) would not make them any easier to sell. One would have to plot a careful course towards a change on such a scale with the introduction potentially staged over time and a comprehensive compensation package put in place. Such a set of policies would equalise implicit carbon taxes for domestic electricity and gas use at between £56/tCO_2e and 59/tCO_2e in 2013–4, very close to the carbon price of £59/tCO_2e that Government estimates as consistent with meeting domestic emissions reduction targets in sectors not covered by the EU ETS.

Table 8.1: Impact of proposed reforms on domestic energy prices, 2013–4

	2013–4 unit price (p/kWh, estimate)	**Effect of 20% VAT rate (p/kWh)**	**Effect of gas tax (p/kWh) including VAT**	**Post-reform unit price (p/kWh)**	**Change (%)**	**Pre-reform carbon price (£/tCO_2e)**	**Post-reform carbon price (£/tCO_2e)**
Electricity	15.60	2.23	0.00	17.83	14.3%	5.92	58.65
Gas	4.83	0.69	0.96	6.47	34.0%	-18.92	56.05

Notes and sources: Pre-reform unit prices are 2012 figures from Department of Energy and Climate Change (2012h for electricity, 2012i for gas) uprated to 2013 values using the year-on-year electricity and gas RPI inflation rates at April 2013. Pre-reform carbon prices are taken from Chapter 6 of this paper.

The efficiency gains from the reforms may be large, given the government commitment to reduce emissions and evidence that the (relatively protected) household sector has previously lagged behind in this regard. The precise impact (on revenue and emissions) will depend on households' price elasticity of demand for energy, which a literature review suggests might be of the order of -0.3 in the short term and -0.8 in the long term. This would imply that the reforms could reduce household emissions by around 8 million tonnes of CO_2 per year in the short run (worth around £400 million per year on current carbon values), and more than 22 million tonnes a year in the long run (worth £1 billion per year), while still generating sufficient revenue to address distributional concerns (Advani et al 2013). This is shown in Table 8.2.

[69] The static revenue raised might be slightly smaller than this if consumers maintain energy consumption by reducing their expenditure on other goods and services that attract the full rate of VAT.

Table 8.2: Revenue and emissions impacts of reforms under different elasticity assumptions

Elasticity	Revenue (£ bn)	Change in CO_2 emissions (million tonnes)			Change as % of 2011 domestic emissions
		Electricity	*Gas*	*Total*	
0	8.2	0.00	0.00	0.00	0.0%
-0.3	7.5	-2.0	-6.4	-8.4	-6.7%
-0.8	6.4	-5.2	-17.1	-22.3	-18.0%
-1.0	5.9	-6.5	-21.4	-27.9	-22.5%

Source: See Advani et al. (2013), Table 6.6.

From an economic efficiency perspective these changes would provide more consistent signals to households to reduce energy use and carbon emissions. Of course the biggest constraint on their implementation is the distributional effect. Their first-order impact (ignoring any behavioural response) would be highly regressive, increasing the cost of living by around 4% for households in the poorest expenditure decile but by less than 1% for those in the richest decile.

However, as demonstrated in the modelling work by Advani et al. (2013) and other recent evidence from Preston et al. (2013), it would be possible to use some of the revenues generated from the tax reforms to design a compensation package which leaves households in the bottom third of the income and expenditure distributions better off on average, and leaving relatively few net losers at the bottom of each distribution. Figures 8.1 and 8.2, taken from Advani et al. (2013), show the combined distributional impact of the pricing reforms and a package of compensatory increases in a range of means-tested benefits.[70] In each figure the top panel shows the distributional effect measured against household income (after housing costs) and the bottom panel the effect measured against non-housing expenditures. Figure 8.1 shows the average effect as a proportion of total income or spending across deciles, and Figure 8.2 the proportion of net winners and losers from the reforms within decile.

The benefit increases cost around £7.2 billion against revenues estimated from the energy reforms of £8.3 billion (perhaps £7.6 billion allowing for short-run behavioural responses in energy demand).

The reform would rationalise carbon pricing, be progressive, and leave some surplus revenue for other expenditure priorities. Although difficult politically, serious consideration should be given to such changes. Our reform package is illustrative, designed to make it clear that a progressive combination of energy prices rises and benefit increases is possible whilst leaving relatively few poor losers overall. Other revenue recycling schemes, for example aimed at the 'squeezed middle' are also possible. Any such change would require a detailed consultation process on the rate and structure of new taxes on domestic gas and non-metered fuels, and the precise compensation package to be implemented not least in order to gain the required buy in

[70] The package also includes the effect of tax thresholds and benefit rates being uprated in line with the one-off increase in the price level of around 1.2 percentage points that the energy price reforms would generate.

Figure 8.1: Average net relative impact of reform package, by decile

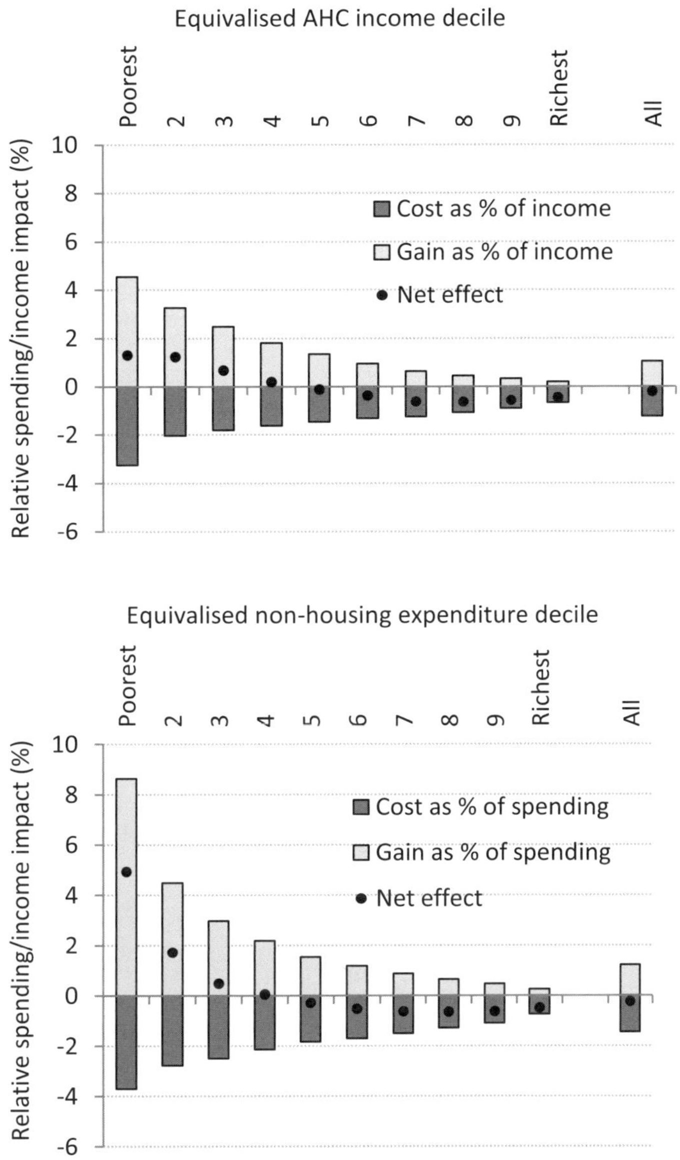

Source: Advani et al. (2013).

The reform modelled above acts is one illustration of how the variation in carbon prices across fuels and end users in 2013 could be reduced. As discussed in Chapter 6, current policy without the VAT subsidy would achieve a carbon price that is approximately equal to the target non-traded carbon price of £59/tCO_2e for electricity and so getting rid of the subsidy gets one close to the 'target' price.

However, in 2020, this will not be the case, with effective carbon prices on electricity far exceeding the target price. Without the VAT subsidy, households will face a carbon price on electricity of £128/tCO_2, twice the stated target rate.

In this situation increasing the VAT rate on energy to 10% would result in an effective carbon price of £65/tCO_2e for households, a very close approximation of the target carbon price in 2020. Such a policy would though leave businesses facing higher prices.

Figure 8.2: Proportion of 'winners' and 'losers' from reform package, by decile

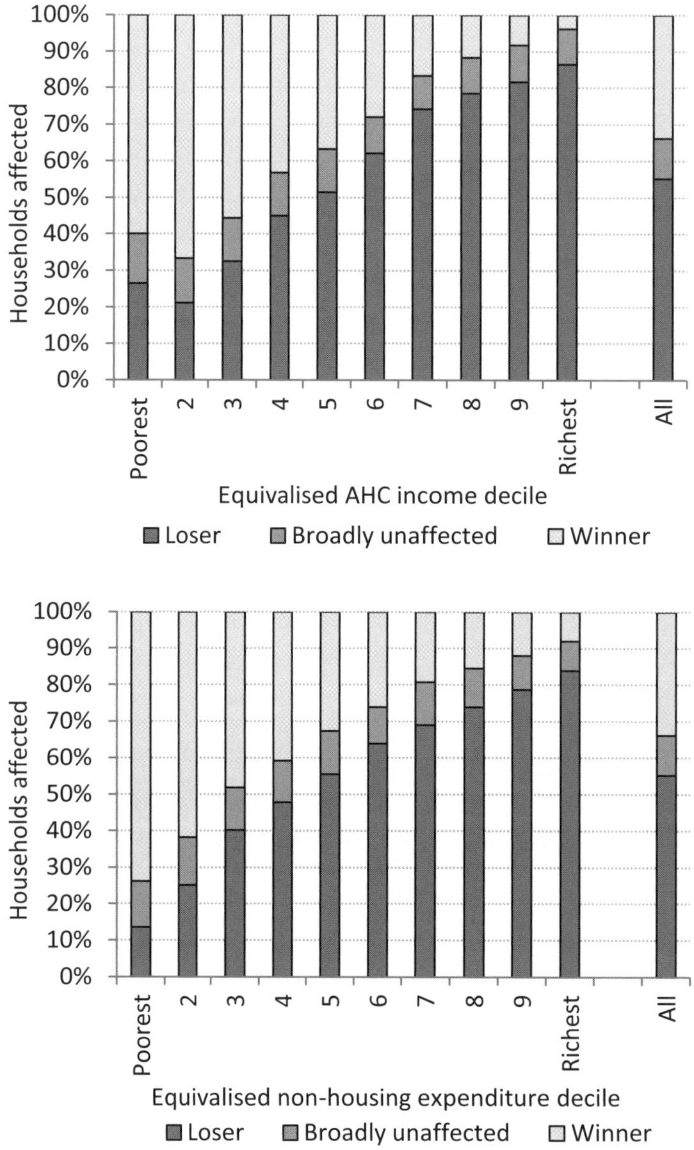

Source: Advani et al. (2013).

In a 'first-best' world, the VAT subsidy on domestic energy use would be removed. Carbon prices could then be reduced by changing other policies. These policies could include those that target bill support (Warm Home Discount), or support for renewables (Renewables Obligation, Contract for Difference FITs) or domestic energy efficiency (ECO). The aims of these policies could be addressed through other methods that do not affect the carbon price. For example, they could be funded through the tax system as opposed to through energy prices.

8.2 Support for small-scale renewable electricity generation

Generous subsidies under small-scale feed-in tariffs (FITs) and relatively rapid declines in the cost of installation led to a much larger-than-anticipated take-up of microgeneration technologies in the household sector, particularly solar PV. Recognising that the desired level of installation could clearly be achieved at lower prices, the government implemented large reductions in the generosity of the scheme several times in 2011 and 2012, with households predictably responding to the incentives to install technologies ahead of pre-announced tariff reductions.

Even after the reductions, though, the scheme continues to be relatively poor value for money in terms of the cost per tonne of carbon saved. At the current generation tariff of 14.9p/kWh for the smallest installations, assuming that FIT generated electricity displaces CCGT as the short-run marginal plant, the carbon price paid for this abatement is very high indeed, making it poor value for money71. Impact assessments for the scheme in 2012 state that, even after the fall in technology costs, the scheme has a negative net present value (around -£4bn), and will impose a cost to consumers of around £700m per year in 2020/21.

In addition, the FIT scheme has also resulted in substantial transfers from the majority of energy consumers towards participants. Wealthier households tend to enrol disproportionately in the scheme. Grover (2013) suggests that this pattern of uptake has led to annual transfers from households in the bottom half of the income distribution to households in the top half each year. This is set to continue until 2020. This suggests that not only does the FIT scheme provide poor value for money in incentivising the uptake of renewable technologies, but also has undesirable distributional consequences.

In the short term, it is difficult to reduce the size of the FIT scheme without creating exactly the kind of uncertainty that could damage future investment prospects. However, one could consider ways to help poorer households benefit. The current existence of higher tariff rates for smaller installations is already suggestive of a desire to spread the gains from this scheme, even at significant efficiency cost. If this is the case, one might be able to improve access for poorer households by making the tariff revenue taxable as with normal income. 'Rent-a-roof' schemes, lending provisions with more competitive interest rates, awareness campaigns, and special provisions for FIT installations at council housing blocks and housing associations could also be considered.

[71] The full carbon price implied by the policy would be £380/tCO_2. This is in fact made up of a resource cost – the actual cost of buying and installing the technology – and a transfer cost – a transfer effectively from electricity consumers to those receiving the FITs. In economic terms the carbon price – the amount we would be willing to spend to reduce emission by1 tonne – should include only the resource cost. If the rest is simply a transfer it is not a cost to the economy of the additional carbon saving, simply a transfer between population groups. Not knowing precisely what in the FIT is the economic cost and what is a transfer we are unable to say what true economic cost of carbon is embedded in the overall tariff.

In the longer run, the rationale for continuing to subsidise small-scale renewables through higher energy costs for other energy consumers will decrease as the microgeneration technologies mature, particularly as we move towards higher carbon price signals. There should be pre-announced reductions in the level of future support which are clearly communicated to potential beneficiaries, moving towards phasing out of small-scale renewable subsidies altogether in the longer-term. This will make it possible for government to send more stable signals to market participants about how the subsidy level will change over time. This should co-ordinate with policy treatment of strike prices for large-scale renewables under the Contract for Difference to ensure that policy does not favour investment in small-scale generation.

Another important lesson from the process of tariff reduction is the importance of modelling accurately the relationship between subsidy level and rate and extent of uptake. Figure 5.1 shows the dramatic effect of the uncertainty created by the "urgent review" of tariff rates. The data from this experience should be used to help better understand the take up decision in the future. This is relevant both in continuing adjustments to small-scale FITs as well as the Renewable Heat Incentive.

8.3 Support for domestic energy-efficiency

Concerns about the distributional effects of higher energy prices along with carbon targets make energy efficiency an obvious goal. The recent announcement of a new fuel poverty target for the efficiency of dwellings for fuel-poor households clearly recognises this, though in general governments should continue to ensure that the financial and non-financial barriers to increased efficiency for *all* households (and businesses) are mitigated.

The flagship Green Deal policy has not had an auspicious beginning. Even allowing for initial teething troubles in establishing the financing mechanisms which have delayed households actually receiving measures, it seems quite possible that the relatively high interest rates likely to be charged and the costs of having the Green Deal assessment will be significant barriers, certainly relative to previous schemes that delivered free or heavily subsidised support for efficiency. These issues make the Green Deal particularly unattractive for relatively inexpensive, easy-to-install measures such as (new or top-up) loft insulation and cavity wall insulation, where credit constraint barriers which are most obviously relaxed by the Green Deal are probably not very significant.

There is therefore an obvious risk that moves towards the Green Deal as the main delivery mechanism for these measures will result in much lower installation rates than even those predicted by the government's own impact assessments. Yet in terms of cost per tonne of carbon saved these measures are extremely worthwhile, paying back extremely quickly. It would therefore be sensible to plan for contingency mechanisms to deliver these measures should take-up under the Green Deal continue to be lower than expected.

Adding additional layers of financial incentives to the Green Deal (such as the current 'cashback' scheme) may ultimately be less efficient than simply financing these measures directly. The total cost of installing remaining loft and easy-to-treat cavity wall insulation is estimated at around £2 billion. Given the relatively modest cost, high impact in terms of reduced carbon and reductions in fuel poverty, it seems sensible to consider as one option direct funding for these measures, ideally delivered alongside the already legislated smart-meter roll-out which sees plans to visit all homes to install new meters within the next few years. Introducing this support alongside the energy price reforms described above would be one sensible way to use any net revenue gains in the first years, helping to improve the efficiency of the domestic housing stock in a cheap, cost-effective way.

A central plank of energy-efficiency policy has been delivering support through obligations on energy suppliers. The current Energy Company Obligation, running alongside the Green Deal, provides direct support to improve insulation and efficiency for people living in hard-to-treat properties or deprived areas.

There is a clear reason to use energy companies as delivery mechanisms for support. However, these policies continue to include relatively prescriptive conditions about what is delivered, to whom, by when and in what form, using different measures to compare delivery (one component of the ECO target is specified in terms of bill savings for vulnerable households, whereas another is in terms of carbon savings in deprived areas, for example). In general, having identified a set of constraints in terms of who is eligible for support and the set of measures which can be delivered to them, it would be preferable to set an overall target for carbon savings and allow energy companies to deliver that target in the most cost-effective way. This could include separate targets for different groups (e.g. a group of poorer households) though it is not clear why targets should be expressed in different terms (carbon or bill savings), and the temptation to prescribe sub-targets should be withheld.[72]

A general problem appears to remain in improving efficiency among private renters. There is evidence that, even controlling for other observable characteristics of the property and the residents, private tenants are much less likely to have insulation measures in their home. The different incentives faced by landlords and tenants are an important market failure and rationalise policy intervention. Under the 2011 Energy Act, landlords will not be able to refuse tenant's requests for measures to be installed with Green Deal financing after 2016, and where after 2018 landlords will no longer be able to rent low-efficiency properties unless all available Green Deal measures have been installed. Government should monitor closely trends in efficiency and insulation in the private rental sector and consider faster moves towards implementing these reforms if progress is slow.

[72] Of course, the ability to change the parameters of how support is delivered should be maintained – for example, the decision to limit the use which energy companies could make of sending free energy-saving light bulbs to consumers under CERT.

8.4 Support for energy bills

A number of existing policies either give direct support to energy bills (the Warm Home Discount), or give labelled cash benefits (Winter Fuel Payment and Cold Weather Payment) which are spent disproportionately on energy and so have at least the flavour of direct bill discounts. In combination, the package of measures strongly favours older households, particularly poorer, older households.

The current Warm Home Discount should set consistent eligibility criteria for the broader group rather than relying on the discretion of individual energy companies to set their own rules, which adds confusion and complexity, could inhibit switching across suppliers, and is characterised at present by a large number of ineligible people receiving a payment.

More generally, in the future the WHD should expand the scope of the core group beyond poorer pensioners to include all eligible households. Automatic payments help prevent stigma issues. Identifying eligible households outside the core group is inhibited because information cannot be shared between DWP and energy companies in the way that is currently possible for Pension Credit.[73] As means-tested benefits are rolled up into Universal Credit (UC) over the next few years, the government should take the opportunity to legislate where necessary to allow receipt of UC to be shared with energy companies in order to identify eligible households automatically.

There are also other potential reforms that could further increase the efficiency of bill support. Winter Fuel Payments are universal payments to individuals over the female state pension age. As their name suggests, the reason given for their introduction was to support pensioners in paying energy bills. But they are in fact just an unhypothecated transfer like any other social security benefit. If the main objective of the policy is to reduce fuel poverty, then it may be sensible to target the payment better on fuel poor households or perhaps in ways which directly support improved energy efficiency. To the extent that this is simply a welfare payment to pensioners that should be made clear and policy decisions taken in the context of broader pension and social security policy.

[73] See Department of Energy and Climate Change (2010), page 31. A recent report by the Energy and Climate Change Select Committee also endorsed wider sharing of data between departments and energy companies (http://www.publications.parliament.uk/pa/cm201314/cmselect/cmenergy/108/10808.htm).

Conclusions

The UK government has set itself targets to reduce GHG emissions. One element of policy aimed at achieving this is, and will continue to be, imposing carbon prices on emissions. This is appropriate, given the stated objective. The correct starting point for such a policy is a single, common carbon price levied on all GHG emissions – which one may then want to vary in specific circumstances, for example if concerned about the impact on some firms engaged in international competition.

Instead, carbon prices in the UK vary substantially across fuels and end users, creating substantial excess costs of abatement. The variation arises from a layering of policies upon policies, differential treatment of different fuels, and differential treatment of different end users. Businesses face higher prices than households, obscuring (but not eliminating) some of the distributional concerns about energy taxation. The effective carbon tax on gas consumption is much lower than that on electricity consumption.

Some variation in the carbon price might be justified by the presence of other objectives. On the household side the government has a fuel poverty target, as well as a potentially more general concern for distribution. Since energy is a necessity, increases in its price are likely to be regressive. For businesses the concern might be that increases in energy prices might lead to carbon leakage, as some processes move offshore to a location where they face a lower tax. If more targeted policies are not possible then non-carbon externalities, the salience of carbon prices, myopic decision-making and dynamic efficiencies, might also provide further rationale for deviations from a single price.

However, even with these other objectives, it is hard to rationalise carbon prices varying by energy type for a particular user. These objectives can also, in most cases, be dealt with directly.

A particularly egregious example of confused policy is the effective subsidy provided to household energy consumption through the imposition of a reduced-rate of VAT. This results in the effective carbon tax on household gas consumption being negative, while that on household electricity consumption is very low. This does not provide the right marginal incentives for emissions reduction. Of course imposing higher taxes would raise energy prices and would be regressive. But there is a strong case for eliminating the VAT subsidy, imposing a tax on gas to ensure that it is treated the same as electricity, and using the revenue raised to deal with the distributional consequences by providing a substantial compensation package for vulnerable households. There are different ways of achieving this and we illustrate one here which protects vulnerable households. Reforms of this type could substantially reduce household emissions, potentially by 22million tonnes of CO_2 a year in the long run (around 4% of current UK emissions on a production basis).

Other household policies, such as those aimed at supporting domestic energy efficiency, are likely to be justified by barriers to change (such as hassle costs), the salience of energy prices, and possible credit constraints. It is important to ensure that these policies are well-targeted to the actual hurdles faced. The Green Deal, for example, mainly helps overcome a lack of access to finance, although the evidence suggests that this is not the most important reason for the lack of take up.

It is important that where the costs of policy are recouped directly from energy bills rather than through taxation, account is taken of the effect this may have on the carbon price. More generally, since it is rarely clear in these circumstances whether the cost of the policy is paid for by households through adjustments to unit price or fixed cost, it would be better to fund such policies directly through taxation. If energy suppliers recover the costs of policy by adjusting fixed costs, this amounts to a household 'poll tax' which is both regressive and fails to provide any carbon reduction incentives.

The same principle of a single carbon price applies to business, but the issue of leakage is more difficult to solve. In the absence of a global agreement on carbon prices, or at least agreement at the EU level, some concession will be necessary to businesses or industries where there is a risk that a high tax in the UK will drive emissions (and economic activity) abroad. More work needs to be done to ensure that the right firms are appropriately protected. There is good evidence that the current Climate Change Agreements – which cover sectors such as craft baking, laundries, and supermarkets – may be too widely drawn and over generous to too many firms. Improvements to targeting could both reduce emissions and provide revenue to the exchequer.

Looking across fuels a key problem appears to be a potential *over*-taxation of electricity. By 2020, upstream prices alone will exceed the non-traded carbon price. This is due to the combination of higher taxes for inputs to non-renewable generation, and support for nuclear and renewables that is recouped through energy prices. Policy needs to be designed such that the way in which these forms of energy are supported is consistent with imposing a uniform carbon price at an appropriate level.

Making good energy policy is not easy because of the number of objectives, many of which are seemingly conflicting. However, even with these myriad targets, the current structure of policy is unnecessarily complex and costly. Incremental reforms are possible that would reduce emissions without harming any of the other objectives, or requiring any new revenue to be committed. Provided they are clearly explained, implementing these reforms should not be politically infeasible.

References

Abeler, J. and Marklein, F. (2010), 'Fungibility, labels and consumption', University of Nottingham Centre for Decision Research and Experimental Economics Discussion Paper 2010-13 (http://www.nottingham.ac.uk/cedex/documents/papers/2010-13.pdf).

Abrell, J., Faye A. N. and Zachmann, G. (2011), 'Assessing the impact of the EU ETS using firm level data', BETA Working Papers 2011–2015. Bureau d'Economie Théorique et Appliquée, UDS, Strasbourg.

Acemoglu, D., Aghion, P. , Bursztyn, L. and Hemous, D. (2012), 'The environment and directed technical change', *American Economic Review*, vol. 102, pp. 131–66 .

Adam, S., Browne, J. and Johnson, P. (2012), 'Pensioners and the tax and benefit system', IFS Briefing Note 130, Institute for Fiscal Studies (http://www.ifs.org.uk/bns/bn130.pdf) .

Advani, A., Leicester, A. and Levell, P. (2011), 'Hyping hypothecation: should green tax revenues be earmarked?', IFS Observation (http://www.ifs.org.uk/publications/5622).

Advani, A., Johnson, P., Leicester, A. and Stoye, G. (2013), 'Household energy use in Britain: a distributional analysis', Institute for Fiscal Studies, Report 85, London: IFS (http://www.ifs.org.uk/comms/r85.pdf)

AEAT (2004), 'Climate change agreements - results of the first target period assessment,. Technical report, AEA Technology, Didcot, UK.

AECOM (2011), 'Energy demand research project: final analysis', AECOM, St. Albans (http://www.ofgem.gov.uk/Sustainability/EDRP/Documents1/Energy%20Demand%20Research%20Project%20Final%20Analysis.pdf).

Aghion, P., Dechezleprêtre, A., Hemous, D., Martin, R. and Van Reenen, J. (2012), 'Carbon taxes, path dependency and directed technical change: evidence from the auto industry', NBER Working Paper no. 18596 (http://www.nber.org/papers/w18596).

Anger, N. and Oberndorfer, U. (2008), 'Firm performance and employment in the EU emissions trading scheme: an empirical assessment for Germany', *Energy Policy*, vol. 36, pp. 12–22.

Atkinson, A. and Stiglitz, J. (1976), 'The design of tax structure: direct versus indirect taxation', *Journal of Public Economics*, vol. 6, pp. 55–75.

Barker, G. (2011), 'Simplifying the CRC Energy Efficiency Scheme – next steps', written ministerial statement by Greg Barker (https://www.gov.uk/government/news/simplifying-the-crc-energy-efficiency-scheme-next-steps-written-ministerial-statement-by-greg-barker).

Barker, T., Ekins, P. and Foxon, T. (2007), 'Macroeconomic effects of efficiency policies for energy-intensive industries: the case of the UK Climate Change Agreements, 2000–2010', *Energy Economics*, vol. 29, pp. 760–78.

Bassi, S., Dechezleprêtre, A. and Fankhauser, S. (2013), 'Climate change policies and the UK business sector'. Grantham Research Institute for Climate Change and the Environment and Centre for Climate Change Economics and Policy, London School of Economics and Political Science.

Beatty, T., Blow, L. and Crossley, T. (2011), 'Is there a "heat or eat"' trade-off in the UK?', Institute for Fiscal Studies, Working Paper no. 11/09 (http://www.ifs.org.uk/wps/wp1109.pdf).

Beatty, T., Blow, L., Crossley, T. and O'Dea, C. (2011), Cash by any other name? Evidence on labelling from the UK Winter Fuel Payment', Institute for Fiscal Studies, Working Paper no. 11/10 (http://www.ifs.org.uk/wps/wp1110.pdf).

Bovenberg, A. and de Mooij, R. (1994), 'Environmental levies and distortionary taxation', *American Economic Review*, vol. 84, pp. 1085–9.

Bovenberg, A. and Goulder, L. (1996) 'Optimal Environmental Taxation in the Presence of Other Taxes: General- Equilibrium Analyses', *American Economic Review*, Vol. 86(4), pp. 985-1000

Bowen, A. (2011), 'The case for carbon pricing', Centre for Climate Change Economics and Policy/Grantham Research Institute on Climate Change Policy Brief (http://www.lse.ac.uk/GranthamInstitute/publications/Policy/docs/PB_case-carbon-pricing_Bowen.pdf).

Bowen, A. and Rydge, J. (2011), 'Climate-change policy in the United Kingdom', Organisation for Economic Co-operation and Development Economics Department Working Paper no. 886 (http://dx.doi.org/10.1787/5kg6qdx6b5q6-en).

Brechling, V. and Smith, S. (1992), 'The pattern of energy efficiency measures amongst domestic households in the UK', Institute for Fiscal Studies Commentary, no. 31.

Browne, J. and Hood, A. (2012), 'A survey of the UK benefits system', Institute for Fiscal Studies Briefing Note no. 13 (http://www.ifs.org.uk/bns/bn13.pdf).

Butler, L. and Neuhoff, K., 2008. 'Comparison of feed-in tariff, quota and auction mechanisms to support wind power development', *Renewable Energy*, vol. 33, pp. 1854–67.

Caird, S., Roy, R. and Herring, H. (2008), 'Improving the energy performance of UK households: results from surveys of consumer adoption and use of low- and zero-carbon technologies', *Energy Efficiency*, vol. 1, pp. 149–66.

Calel, R. (2013). 'Emissions trading and technological change'. PhD Thesis, Department of Geography and the Environment, London School of Economics.

Carbon Trust (2010), 'Tackling carbon leakage – sector-specific solutions for a world of unequal carbon prices' (http://www.carbontrust.com/media/84908/ctc767-tackling-carbon-leakage.pdf).

Carbon Trust (2012), 'CRC energy efficiency scheme' (http://www.carbontrust.com/resources/guides/carbon-footprinting-and-reporting/crc-carbon-reduction-commitment).

Carbon Trust and SPA Future Thinking (2012), 'Exploring the design of policies to increase efficiency of electricity use within the industrial and commercial sectors', DECC Research Report (https://www.gov.uk/government/uploads/system/uploads/attachment_data/file/66566/7028-design-policies-efficiency-elec-edr.pdf)

Centre for Strategy and Evaluation Services and Oxford Research (2012), 'Evaluation of the Ecodesign Directive (2009/125/EC) Final Report' (http://ec.europa.eu/enterprise/policies/sustainable-business/ecodesign/review/index_en.htm)

Chan, H. S., Li, S. and Zhang, F. (2012) 'Firm competitiveness and the European Union Emissions Trading Scheme', World Bank Working Paper.

Cherp, A. and Jewell, J. (2011), 'The three perspectives on energy security: intellectual history, disciplinary roots and the potential for integration', *Current Opinion in Environmental Sustainability*, vol. 3, pp. 202–12.

Chetty, R., Looney, A. and Kroft, K. (2009), 'Salience and taxation: theory and evidence', *American Economic Review*, vol. 99, pp. 1145–77.

Chevallier, J. (2011), 'Carbon price drivers: an updated literature review'. Working Paper (http://ssrn.com/abstract=1811963 or http://dx.doi.org/10.2139/ssrn.1811963).

Chitnis, M., Sorrell, S., Druckman, A., Firth, S. and Jackson, T. (2013), 'Turning lights into flights: estimating direct and indirect rebound effects for UK households', *Energy Policy*, vol. 55,pp. 234–50.

Cobb, M., Kenber, M. and Haugen, O. (2009), 'The effects of EU climate legislation on business competitiveness : a survey and analysis', Climate and Energy Paper Series, no. 9, The Climate Group.

Committee on Climate Change (2010), 'Building a low-carbon economy – the UK's innovation challenge'. Supporting analysis and review of evidence.

(http://downloads.theccc.org.uk.s3.amazonaws.com/Low%20carbon%20Innovation/Low carbon innovation supporting-analysis.pdf).

Committee on Climate Change (2012), 'Public letter from Lord Adair Turner to the Secretary of State. 27 March 2012'. (http://downloads.theccc.org.uk.s3.amazonaws.com/Letters/EdwardDavey MP Letter270312.pdf).

Committee on Climate Change (2013a), 'Reducing the UK's carbon footprint and managing competitiveness risks'(http://www.theccc.org.uk/publication/carbon-footprint-and-competitiveness/).

Committee on Climate Change (2013b), 'Meeting Carbon Budgets – 2013 Progress Report to Parliament'.(http://www.theccc.org.uk/wp-content/uploads/2013/06/CCC-Prog-Rep-Book singles web 1.pdf).

Committee on Climate Change (2013c), 'Next steps on electricity market reform – securing the benefits of low-carbon investment' (http://www.theccc.org.uk/publication/next-steps-on-electricity -market-reform-23-may-2013/).

Cramton, P. and Kerr, S. (2002), 'Tradeable carbon permit auctions – how and why to auction not grandfather', *Energy Policy*, vol. 30, pp. 333–45.

Crossley, T. and O'Dea, C. (2010), 'How to keep warm in winter: Winter Fuel Payments or Cold Weather Payments?', Institute for Fiscal Studies Observation (http://www.ifs.org.uk/publications/5308).

De Bruyn, S.,Markowska, A. and Nelissen, D. (2010), 'Will the energy-intensive industry profit from EU ETS under Phase 3?', Report, CE Delft.

De Bruyn, S., Nelissen, D. and Koopman, M. (2013), 'Carbon leakage and the future of the EU ETS market – impact of recent developments in the EU ETS on the list of sectors deemed to be exposed to carbon leakage', Report, CE Delft (http://www.cedelft.eu/art/uploads/CE Delft 7917 Carbon leakage future EU ETS market Final.pdf).

Department of Energy and Climate Change and Energy Saving Trust (2011), 'Home energy pay as you save pilot review' (http://www.energysavingtrust.org.uk/content/download/3984/77555/versio/1/file/PAYS+Review+Report+Final+05 09 11.pdf).

Department of Energy and Climate Change and HM Treasury (2013a), Background documentation, in `Green Book supplementary appraisal guidance on valuing energy use and greenhouse gas emissions' , Chapter 2 (https://www.gov.uk/government/uploads/system/uploads/attachment data/file/243939/2013 Appraisal Guidance - Background documentation - FINAL.pdf).

Department of Energy and Climate Change and HM Treasury (2013b) 'Green Book supplementary appraisal guidance on valuing energy use and greenhouse gas emissions', (https://www.gov.uk/government/uploads/system/uploads/attachment_data/file/68944/Tables_1-20.xlsx)

DEFRA and DECC (2012), '2012 Guidelines to Defra/DECC's GHG Conversion Factors for Company Reporting', Department for Environment, Food and Rural Affairs (http://www.defra.gov.uk/publications/files/pb13773-ghg-conversion-factors-2012.pdf).

DellaVigna, S. (2009), 'Psychology and economics: evidence from the field', *Journal of Economic Literature*, vol. 47, pp. 315–72.

Department for Business, Innovation and Skills and Department of Energy and Climate Change (2012). Energy intensive industries in the UK – maintaining international competitiveness. Compensation for the indirect costs of EU ETS and Carbon Price Support, Consultation on scheme eligibility and design. (https://www.gov.uk/government/uploads/system/uploads/attachment_data/file/32707/12-1179-energy-intensive-industries-compensation-consultation-on-scheme.pdf).

Department for Environment, Food and Rural Affairs (2009), 'Saving energy through better products and appliances' (http://archive.defra.gov.uk/environment/economy/documents/energy-products-1209.pdf).

Department of Energy and Climate Change (2009), 'Carbon valuation in UK policy appraisal: a revised approach', Report (https://www.gov.uk/government/uploads/system/uploads/attachment_data/file/41798/1_20090715105804_e____carbonvaluationinukpolicyappraisal.pdf).

Department of Energy and Climate Change (2010), 'Consultation on the Warm Home Discount' (https://www.gov.uk/government/uploads/system/uploads/attachment_data/file/42589/956-consultation-warm-home-discount.pdf).

Department of Energy and Climate Change (2011a), 'The warm home discount scheme impact assessment', Report (https://www.gov.uk/government/uploads/system/uploads/attachment_data/file/42595/1308-warm-home-disc-impact-assessment.pdf).

Department of Energy and Climate Change (2011b), 'Climate Change Agreements: results of the fifth target period' (https://www.gov.uk/government/uploads/system/uploads/attachment_data/file/49000/3366-cca-5th-target.pdf).

Department of Energy and Climate Change (2011c), 'Electricity market reform: update on the emissions performance standard'. London: DECC. (https://www.gov.uk/government/uploads/system/uploads/attachment_data/file/48375/5350-emr-annex-d--update-on-the-emissions-performance-s.pdf).

Department of Energy and Climate Change (2011d),'Estimated impact of energy and climate change policies on energy prices and bills' (https://www.gov.uk/government/uploads/ system/uploads/attachment data/file/68820/3593-estimated-impacts-of-our-policies-on-energy-prices.pdf).

Department of Energy and Climate Change (2011e),'Renewable heat incentive' (https://www.gov.uk/government/uploads/system/uploads/attachment data/file/48041/1387-renewable-heat-incentive.pdf).

Department of Energy and Climate Change (2012a), 'Impact assessment: government response to the consultation on the proposals for the levels of banded support under the Renewables Obligation for the period 2013-17 and the Renewables Obligation Order 2012' (https://www.gov.uk/government/uploads/system/uploads/attachment data/file/42847/5945-renewables-obligation-government-response-impact-a.pdf).

Department of Energy and Climate Change (2012b), 'Impact assessment: government response to consultation on the Comprehensive Review Phase 2B – on Feed-In Tariffs for anaerobic digestion, wind, hydro and micro-CHP installations' (https://www.gov.uk/government/ uploads/system/uploads/attachment data/file/42910/5903-impact-assessment-govt-response-to-consultation-f.pdf).

Department of Energy and Climate Change (2012c), 'Electricity Market Reform: Policy Overview' ((https://www.gov.uk/government/uploads/system/uploads/attachment data/file /65634/7090-electricity-market-reform-policy-overview-.pdf).

Department of Energy and Climate Change (2012d), 'Consultation on simplifying the CRC Energy Efficiency Scheme' (https://www.gov.uk/government/ uploads/system/uploads/attachment data/file/42931/4757-cons-simp -crc-energy-efficiency-scheme.pdf).

Department of Energy and Climate Change (2012e), 'Consultation on simplifying the CRC Energy Efficiency Scheme – government response' (https://www.gov.uk/government/uploads/system/uploads/attachment data/file/68946/Consultation on Simplifying the CRC Energy Efficiency Scheme - Government Response.pdf).

Department of Energy and Climate Change (2012f), 'Estimates of Home Insulation Levels in Great Britain: April 2013', (https://www.gov.uk/government/uploads/system/uploads/attachment_data/file/209029/Statistical_release_-_Estimates_of_home_insulation_levels_in_Great_Britain_April_13.pdf).

Department of Energy and Climate Change (2012g), 'Final stage impact assessment for the Green Deal and Energy Company Obligation' (https://www.gov.uk/government/uploads/system/

uploads/attachment_data/file/42984/5533-final-stage-impact-assessment-for-the-green-deal-a.pdf).

Department of Energy and Climate Change (2012h), 'Average variable unit costs and fixed costs for electricity for selected towns and cities in the UK', Report (https://www.gov.uk/government/uploads/system/uploads/attachment_data/file/172856/qep224.xls).

Department of Energy and Climate Change (2012i), 'Average variable unit costs and standing charges for gas in 2012 for selected towns and cities in Great Britain', Report (https://www.gov.uk/government/uploads/system/uploads/attachment_data/file/172860/qep234.xls).

Department of Energy and Climate Change (2012j), 'Emissions performance standard impact assessments (IAs)', May (https://www.gov.uk/government/uploads/system/uploads/attachment_data/file/204801/eps_ia.pdf).

Department of Energy and Climate Change, (2012k), 'Capacity market: design and implementation update', November (https://www.gov.uk/government/uploads/system/uploads/attachment_data/file/65637/7104-emr-annex-c-capacity-market-design-and-implementat.pdf).

Department of Energy and Climate Change (2012l), 'Smart metering implementation programme smart metering equipment technical specifications' (https://www.gov.uk/government/uploads/system/uploads/attachment_data/file/43087/6425-smart-metering-equipment-technical-specifications-.pdf).

Department of Energy and Climate Change (2012m), 'Smart meter roll-out for the non-domestic sector (GB) April 2012' (https://www.gov.uk/government/uploads/system/uploads/attachment_data/file/48805/4907-smart-meter-rollout-non-domestic-ia-resp.pdf).

Department of Energy and Climate Change (2012n), 'Updated short-term traded carbon values used for UK public policy appraisal', October 2012. (https://www.gov.uk/government/uploads/system/uploads/attachment_data/file/245385/6667-update-short-term-traded-carbon-values-for-uk-publ.pdf).

Department of Energy and Climate Change (2012o), *Digest of UK Energy Statistics*, London: DECC (https://www.gov.uk/government/organisations/department-of-energy-climate-change/series/digest-of-uk-energy-statistics-dukes).

Department of Energy and Climate Change (2013a), 'Annual report on fuel poverty statistics 2013' (https://www.gov.uk/government/uploads/system/uploads/attachment_data/file/199833/Fuel_Poverty_Report_2013_FINALv2.pdf).

Department of Energy and Climate Change (2013b). 'Estimated impacts of energy and climate change policies on energy prices and bills' (https://www.gov.uk/government/uploads/system/uploads/attachment_data/file/172923/130326_-_Price_and_Bill_Impacts_Report_Final.pdf).

Department of Energy and Climate Change (2013c), '2012 EU ETS results – UK sector table' (https://www.gov.uk/government/uploads/system/uploads/attachment_data/file/181586/2012_eu_ets_results_april_2013.pdf).

Department of Energy and Climate Change (2013d), '2012 UK greenhouse gas emissions, provisional figures, and 2011 UK greenhouse gas emissions, final figures by fuel type and end-user', (https://www.gov.uk/government/uploads/system/uploads/attachment_data/file/193414/280313_ghg_national_statistics_release_2012_provisional.pdf).

Department of Energy and Climate Change (2013e), 'Commons debate – written answers (electricity generation) 3 June 2013' (http://www.publications.parliament.uk/pa/cm201314/cmhansrd/cm130603/text/130603w0002.htm#130603w0002.htm_wqn54).

Department of Energy and Climate Change (2013f), *Digest of United Kingdom energy statistics (DUKES) 2013.* (https://www.gov.uk/government/publications/digest-of-united-kingdom-energy-statistics-dukes-2013-printed-version-excluding-cover-pages).

Department of Energy and Climate Change (2013g), 'Participating in the EU ETS' (https://www.gov.uk/participating-in-the-eu-ets).

Department of Energy and Climate Change, (2013h), 'Climate change levy' (http://customs.hmrc.gov.uk/channelsPortalWebApp/channelsPortalWebApp.portal?_nfpb=true&_pageLabel=pageExcise_InfoGuides&propertyType=document&id=HMCE_CL_001174).

Department of Energy and Climate Change, (2013i), 'Climate Change Agreements'. (https://www.gov.uk/climate-change-agreements).

Department of Energy and Climate Change (2013j), 'Simplification options for the CRC Energy Efficiency scheme to help business: CRC (Amendment) Order 2013', Impact Assessment (IA) no. DECC0066.

Department of Energy and Climate Change (2013k) 'Electricity market reform: delivering UK investment' (https://www.gov.uk/government/uploads/system/uploads/attachment_data/file/209276/EMR_Spending_Review_Announcement_-_FINAL_PDF.pdf).

Department of Energy and Climate Change (2013l), 'Electricity market reform: ensuring electricity security of supply and promoting investment in low-carbon generation', Impact Assessment (IA) no. DECC0130.

Department of Energy and Climate Change (2013m), 'Tariffs and technologies affected by the 2013 Non Domestic Early Tariff Review' (https://www.gov.uk/government/uploads/system/uploads/attachment_data/file/204450/Tariffs_and_technologies_affected_by_the_2013_Tariff_Review_3.pdf)

Department of Energy and Climate Change (2013n), 'Domestic Green Deal and Energy Company Obligation in Great Britain', Monthly Report (May 2013) (https://www.gov.uk/government/uploads/system/uploads/attachment_data/file/198715/statistical_release_green_deal_eco_may_2013.pdf).

Department of Energy and Climate Change (2013o), 'Smart Meters Programme: Smart Meters Programme Delivery Plan' (https://www.gov.uk/government/uploads/system/uploads/attachment_data/file/197794/smart_meters_programme.pdf).

Department of Energy and Climate Change (2013p), 'Smart Meter roll-out for the domestic and small and medium non-domestic sectors (GB)' (https://www.gov.uk/government/uploads/system/uploads/attachment_data/file/78666/IA-Feb.pdf)

Department for Trade and Industry (2007), 'Meeting the Energy Challenge – A White Paper on Energy' (http://www.berr.gov.uk/files/file39387.pdf).

Department of Work and Pensions (2011), 'Outcome of the energy rebate scheme' (https://www.gov.uk/government/uploads/system/uploads/attachment_data/file/174967/outcome-energy-rebate-scheme.pdf.pdf).

Department of Work and Pensions (2012), 'income related benefits: estimates of take-up in 2009–10' (http://research.dwp.gov.uk/asd/income_analysis/feb2012/tkup_full_report_0910.pdf).

Dröge, S. (2013), 'Carbon pricing and its future role for energy-intensive industries – key features of steel, cement, aluminium, basic chemicals, pulp and paper', Climate Strategies, London (http://www.climatestrategies.org/research/our-reports/category/61/367.html).

Dröge, S. and Cooper, S. (2010), 'Tackling leakage in a world of unequal carbon prices – a study for the Greens/EFA Group', Climate Strategy, London (http://www.climatestrategies.org/research/our-reports/category/61/257.html).

Ekins, P. and Etheridge, B. (2006), 'The environmental and economic impacts of the UK Climate Change Agreements'. *Energy Policy*, vol. 34, pp. 2071–86.

Ekins, P. (2010), 'A major green fiscal reform for the UK: results for the economy, employment and the environment', Green Fiscal Commission Briefing Paper 5,

Policy Studies Institute, London (http://www.greenfiscalcommission.org.uk/images/uploads/gfcBriefing5_PDF_isbn_v5.pdf).

Environmental Audit Committee (2008) 'Reducing Carbon Emissions from UK Business: The role of the Climate Change Levy and Agreements Second Report of Session 2007–08'. London: EAC. (http://www.publications.parliament.uk/pa/cm200708/cmselect/cmenvaud/354/354.pdf).

Ellerman, A. D. and Buchner, B. (2008), 'Over-allocation or abatement? A preliminary analysis of the EU ETS based on the 2005–06 emissions data', *Environmental and Resource Economics*, vol.. 1, pp. 267–87.

Ellison, G. and Ellison, S. (2009), 'Search, obfuscation and price elasticities on the Internet', *Econometrica*, vol. 77, pp. 111–32.

Environment Agency (2013), 'CRC Energy Efficiency Scheme guidance for participants in Phase 1 (2010–2011 to 2013–2014)' (http://a0768b4a8a31e106d8b0-50dc802554eb38a24458b98ff72d550b.r19.cf3.rackcdn.com/LIT_6794_b04f9e.pdf).

Environmental Audit Committee (2008), 'Reducing Carbon Emissions from UK Business: The role of the Climate Change Levy and Agreements Second Report of Session 2007–08'. London: EAC. (http://www.publications.parliament.uk/pa/cm200708/cmselect/cmenvaud/354/354.pdf)

Espey, J. and Espey, M. (2004), 'Turning on the lights: a meta-analysis of residential electricity demand elasticities', *Journal of Agricultural and Applied Economics*, vol. 36, pp. 65–81.

European Commission (2005), 'The support of renewable energy sources'. European Commission Report COM (2005) 627.

European Commission (2009), 'Directive 2009/29/EC of the European Parliament and of the Council of 23 April 2009 amending Directive 2003/87/EC so as to improve and extend the greenhouse gas emission allowance trading scheme of the Community', L140/63 OJ 5.6.2009 (http://eur-lex.europa.eu/LexUriServ/LexUriServ.do?uri=OJ:L:2009:140:0063:0087:EN:PDF).

European Commission (2012a), 'Commission staff working document: establishment of the Working Plan 2012–2014 under the Ecodesign Directive' (http://ec.europa.eu/enterprise/policies/sustainable-business/documents/eco-design/working-plan/files/comm-swd-2012-434-ecodesign_en.pdf).

European Commission (2012b), 'Ecodesign your future' (http://ec.europa.eu/enterprise/policies/sustainable-business/ecodesign/files/brochure_ecodesign_en.pdf).

European Commission (2013), 'The EU Emissions Trading System (EU ETS)'. (http://ec.europa.eu/clima/publications/docs/factsheet_ets_2013_en.pdf).

European Environment Agency (2013), 'EU Emissions Trading System (ETS) data viewer'. (http://www.eea.europa.eu/data-and-maps/data/data-viewers/emissions-trading-viewer).

Fankhauser, S., Hepburn, C. and Park, J. (2010),. 'Combining multiple climate policy instruments: how not to do it', *Climate Change Economics*, vol. 1, pp. 209–25.

Friedman, L. (2002), 'Bounded rationality versus standard utility-maximization: a test of energy price responsiveness', in Gowda, R. and J. Cox (eds.), *Judgments, Decisions and Public Policy*, Cambridge: Cambridge University Press.

Fullerton, D., Leicester, A. and Smith, S. (2010), 'Environmental taxes', in J. Mirrlees, S. Adam, T. Besley, R. Blundell, S. Bond, R. Chote, M. Gammie, P. Johnson, G. Myles and J. Poterba (eds), *Dimensions of Tax Design: the Mirrlees Review*, Oxford: Oxford University Press.

Geman, H. (2005), *Commodities and commodity derivatives*. Chichester: John Wiley and Sons.

Gigerenzer, G. and Brighton, H. (2009), 'Homo heuristicus: why biased minds make better inferences', *Topics in Cognitive Science I*, vol. 1, pp. 107–43.

Gillingham, K. and Sweeney, J. (2010), 'Market failure and the structure of externalities', Yale University, mimeo (http://www.yale.edu/gillingham/Market%20Failure%20and%20the%20Structure%20of%20Externalities.pdf).

Gillingham, K., Newell, R. and Palmer, K. (2009), 'Energy efficiency economics and policy', Resources for the Future Discussion Paper 09-13 (http://www.rff.org/documents/rff-dp-09-13.pdf).

Gouldson, A. (2004), 'Risk, regulation and the right to know: exploring the impacts of access to information on the governance of environmental risk', *Sustainable Development*, vol. 12, pp. 136–49.

Graichen, V., Schumacher, K, Matthes, F. C. Mohr, L., Duscha, V., Schleich, J. and Diekmann , J. (2008), 'Impacts of the EU Emissions Trading Scheme on the industrial competitiveness in Germany: final draft report prepared under the UFOPLAN'. Dessau-Roßlau: Federal Environment Agency (Umweltbundesamt) (http://www.umweltbundesamt.de/sites/default/files/medien/publikation/long/3625.pdf).

Greene, D. (2010), 'How consumers value fuel economy: a literature review', US Environmental Protection Agency (http://www.epa.gov/otaq/climate/regulations/420r10008.pdf).

Grover, D. (2013), 'The British Feed-in Tariff for small renewable energy systems: Can it be made fairer?', Grantham Research Institute for Climate Change and the Environment and Centre for Climate Change Economics and Policy, London School of Economics and Political Science.

Grubb, M. and Counsell, T. (2009), 'Tackling carbon leakage – sector-specific solutions for a world of unequal carbon prices', Carbon Trust Working Paper (http://www.carbontrust.com/media/84908/ctc767-tackling-carbon-leakage.pdf).

Gruell, G. and Taschini, L. (2012), 'A Comparison of Reduced-Form Permit Price Models and Their Empirical Performances', *Proceedings of the 58th World Statistics Congress*. Dublin, Ireland.

Hassett, K. and Metcalfe, G. (1993), 'Energy conservation investment: do consumers discount the future correctly?', *Energy Policy*, vol. 21, pp. 710–6.

Hausman, J. (1979), 'Individual discount rates and the purchase and utilization of energy-using durables', *The Bell Journal of Economics*, vol. 10, pp. 33–54.

Helm, D. (2010), 'Government failure, rent-seeking, and capture: the design of climate-change policy', *Oxford Review of Economic Policy*, vol. 26, issue 2, pp. 182–96.

Helm, D. (2012), 'EMR and the Energy Bill: a critique' (http://www.dieterhelm.co.uk/node/1330).

Helm, D., Hepburn, C. and Ruta, G. (2012). 'Trade, climate change and the political game theory of border carbon adjustments', Working Paper no. 92, Centre for Climate Change Economics and Policy and Grantham Research Institute on Climate Change, London School of Economics.

Hepburn, C, Grubb, M., Neuhoff, K., Matthes F. and Tse, M. (2013), 'Auctioning of EU ETS phase II allowances: how and why?' *Climate Policy,* vol 6, issue no. 2006, pp. 137–60.

Heutel, G. (2011), 'Optimal policy instruments for externality-producing durable goods under time inconsistency', NBER Working Paper no. 17083 (http://www.uncg.edu/bae/people/heutel/Papers/Heutel_PresentBias.pdf).

Hills, J. (2011), 'Fuel poverty – the problem and its measurement: interim report of the Fuel Poverty Review', CASE Report 69 (http://sticerd.lse.ac.uk/dps/case/cr/CASEreport69_Full_Report.pdf).

Hills, J. (2012), 'Getting the measure of fuel poverty – final report of the Fuel Poverty Review', CASE Report 72 (http://sticerd.lse.ac.uk/dps/case/cr/CASEreport72.pdf).

Hirst, E. and Brown, M. (1990), 'Closing the efficiency gap: barriers to the efficient use of energy', *Resources, Conservation and Recycling*, vol 3, pp. 267–81.

Hitzemann, S. and Uhrig-Homburg, S. (2013), 'Understanding the price dynamics of emission permits: a model for multiple trading periods', Working Paper, Universität Karlsruhe (TH). (http://papers.ssrn.com/sol3/papers.cfm?abstract_id=1763182).

HM Revenue and Customs (2012a), 'Main tax expenditures and structural reliefs' (http://www.hmrc.gov.uk/statistics/expenditures/table1-5.pdf).

HM Revenue and Customs (2012b), 'Notice CCL1/3 Climate Change Levy – reliefs and special treatments for taxable commodities', May (http://customs.hmrc.gov.uk/channelsPortalWebApp/downloadFile?contentID=HMCE_CL_000181).

HM Revenue and Customs (2013a.), 'HMRC Tax & NIC Receipt May 2013' (http://www.hmrc.gov.uk/statistics/receipts/info-analysis.pdf).

HM Revenue and Customs (2013b), 'Climate Change Agreements scheme - industrial and commercial sectors 2013 to 2023' (http://www.hmrc.gov.uk/climate-change-levy/industrial-commercial.pdf).

HM Revenue and Customs (2013c), 'Climate Change Agreements scheme – agricultural sectors 2013 to 2023', HM Revenue and Customs (http://www.hmrc.gov.uk/climate-change-levy/agricultural-sectors.pdf).

HM Treasury (2010), 'Spending Review 2010' (https://www.gov.uk/government/uploads/system/uploads/attachment_data/file/203826/Spending_review_2010.pdf).

HM Treasury (2011a), 'Carbon price floor consultation: the government response' (https://www.gov.uk/government/uploads/system/uploads/attachment_data/file/190279/carbon_price_floor_consultation_govt_response.pdf).

HM Treasury (2011b), 'Control framework for DECC levy-funded spending' (http://webarchive.nationalarchives.gov.uk/20130129110402/http://www.hm-treasury.gov.uk/d/control_framework_decc250311.pdf).HM Treasury (2012), 'Autumn Statement 2012' (https://www.gov.uk/government/uploads/system/uploads/attachment_data/file/221550/autumn_statement_2012_complete.pdf).

HM Treasury (2013), 'Written Ministerial Statement – environmental taxation' (http://www.parliament.uk/documents/commons-vote-office/March-2013/26-3-13/2.CHANCELLOR-Environmental-taxation.pdf).

HM Treasury and HM Revenues and Customs (2010a), 'Impact assessment of proposals to amend the Climate Change Levy and fuel duty to support incentives for low-carbon electricity generation' (https://www.gov.uk/government/uploads/system/uploads/attachment_data/file/81274/consult_carbon_price_support_ia.pdf).

HM Treasury and HM Revenues and Customs (2010b), 'Carbon price floor: support and certainty for low-carbon investment' (https://www.gov.uk/government/uploads/system/uploads/attachment_data/file/42639/consult_carbon_price_support_condoc.pdf).

Horn, H. and Mavroidis, P. (2011), 'To B(TA) or not to B(TA)? On the legality and desirability of border tax adjustments from a trade perspective', *The World Economy*, vol. 34, pp. 1911–37.

Hough, D. and Bolton, P. (2012), 'The Warm Home Discount Scheme', House of Commons Library Standard Note SN/SC 05956 (http://www.parliament.uk/briefing-papers/SN05956.pdf).

Hourcade, J.-C., Demailly, D. , Neuhoff, K. and Sato, M. (2007), 'Differentiation and dynamics of EU ETS industrial competitiveness impact'. London: Climate Strategies.

House of Commons (2009), 'Climate Change Levy', Parliamentary Briefing Paper (http://www.parliament.uk/briefing-papers/SN00235.pdf).

House of Commons (2012a), 'Draft Energy Bill: pre-legislative scrutiny. First Report of Session 2012–13' (http://www.publications.parliament.uk/pa/cm201213/cmselect/cmenergy/275/275.pdf).

House of Commons (2012b), 'The renewables obligation'. Standard Note: SN/SC/5870 (http://www.parliament.uk/briefing-papers/sn05870.pdf).

House of Commons (2012c), 'Community Energy Saving Programme (CESP)' Standard Note no. SN/SC/06197 (http://www.parliament.uk/briefing-papers/SN06197.pdf%E2%80%8E).

House of Commons Energy and Climate Change Committee (2013), *Energy Prices, Profits and Poverty*, London: The Stationery Office Limited (http://www.publications.parliament.uk/pa/cm201314/cmselect/cmenergy/108/108.pdf).

Huhne, C. (2010), 'The Rt Hon Chris Huhne MP's speech to the CBI Conference' (https://www.gov.uk/government/speeches/the-rt-hon-chris-huhne-mps-speech-to-the-cbi-conference).

Institute for Public Policy Research (2011), 'Hot air: the carbon floor price in the UK' (http://ippr.org/publication/55/7629/hot-air-the-carbon-price-floor-in-the-uk).

IPA Energy Consulting (2005), 'Implication of the EU Emissions Trading Scheme for the UK power generation sector', Department for Trade and Industry (http://www.ipaeconomics.com/images/cmsimages/media/downloads/general/2005/FINAL%20Report%201867%2011-11-05.pdf).

IPCC (1999), *Aviation and the Global Atmosphere*, J. E. Penner, D. H. Lister, D. J. Griggs, D. J. Dokken, M. McFarland (eds). Cambridge: Cambridge University Press. (http://www.ipcc.ch/ipccreports/sres/aviation/)

Ismer, R. and Neuhoff, K. (2007), 'Border tax adjustment: a feasible way to support stringent emission trading', *European Journal of Law and Economics*, vol. 24, pp. 137–64.

Iyengar, S. and Lepper, M. (2000), 'When choice is demotivating: can one desire too much of a good thing?', *Journal of Personality and Social Psychology*, vol. 79, pp. 995–1006.

Jaffe, A. and Stavins, R. (1994), 'The energy efficiency gap: what does it mean?', *Energy Policy*, vol. 22, pp. 804–10.

Jaffe, A., Newell, R. and Stavins, R. (2005), 'A tale of two market failures: technology and environmental policy', *Ecological Economics*, vol. 54, 164–74.

Johnson, P., Leicester, A. and Stoye, G. (2012), *Fuel for Thought: the What, Why and How of Motoring Taxation*, London: RAC Foundation (http://www.racfoundation.org/assets/rac_foundation/content/downloadables/fuel_for_thought-johnson_et_al-150512.pdf).

Lacombe, R. H. (2008). 'Economic impact of the European union emission trading scheme : evidence from the refining sector', MTech Thesis, Massachusetts Institute of Technology.

Leicester, A. (2006), 'The UK tax system and the environment', IFS Report no. 68 (http://www.ifs.org.uk/comms/r68.pdf).

Leicester, A. and Stoye, G. (2012),' A defining Issue? The government's pledge to raise the share of revenue from green taxes', Institute for Fiscal Studies Observation (http://www.ifs.org.uk/publications/6491).

Leicester, A. and Stoye, G. (2013), 'People or places? Factors associated with the presence of domestic energy efficiency measures in England', Institute for Fiscal Studies, Working Paper no. W13/14 (http://www.ifs.org.uk/wps/wp201314.pdf).

Mansanet-Bataller, M., Pardo A. and Valor, E. (2007). 'CO_2 prices, energy and weather', *The Energy Journal*, vol. 28, issue 3, pp. 67–86.

Martin, R. and Wagner U. J. (2009), 'Climate change policy and innovation', Centre for Economic Performance (http://gcoe.ier.hit-u.ac.jp/CAED/papers/id115_Martin_Wagner.pdf).

Martin, R., de Preux, L. B. and Wagner, U. J. (2011), 'The impacts of the Climate Change Levy on manufacturing: evidence from microdata',National Bureau of Economic Research Working Paper Series, 17446 (http://ssrn.com/abstract=1933049).

Martin, R., Muûls, M. and Wagner, U. J. (2012), *An Evidence Review of the EU Emissions Trading System, Focussing on Effectiveness of the System in Driving Industrial Abatement.* London: DECC. (https://www.gov.uk/government/uploads/system/uploads/attachment_data/file/48449/5725-an-evidence-review-of-the-eu-emissions-trading-sys.pdf).

McKinsey and Company (2012), *Capturing the Full Electricity Efficiency Potential of the UK*, London: DECC (https://www.gov.uk/government/uploads/system/uploads/attachment_data/file/65626/7035-capturing-full-elec-eff-potential-edr.pdf).

McKinsey and Company, and Ecofys (2006). 'EU ETS Review: report on international competitiveness'.

Mirrlees, J., Adam, S., Besley, T., Blundell, R., Bond, S., Chote, R., Gammie, M., Johnson, P., Myles, G. and Poterba, J. (2011), *Tax by Design: the Mirrlees Review*. Oxford: Oxford University Press (http://www.ifs.org.uk/mirrleesReview/design).

Mitchell, C. and Connor, P. M. (2004), 'Renewable energy policy in the UK 1990–2003', *Energy Policy*, vol. 32, pp. 1935–47.

Mitchell, C., Bauknecht, D. and Connor, P. M. (2006), 'Effectiveness through risk reduction: a comparison of the renewable obligation in England and Wales and the feed-in system in Germany'. *Energy Policy*, vol. 34, pp. 297–305.

Monjon, S. and Quirion, P. (2009), 'Addressing leakage in the EU ETS: results from the CASE II model', Climate Strategies Working Paper, no. 3. (http://www.climatestrategies.org/component/reports/category/32/141.html).

National Audit office (2007), 'The Climate Change Levy and Climate Change Agreements' (http://www.nao.org.uk/wp-content/uploads/2012/11/climate_change_review.pdf).

National Audit Office (2011), *Preparations for the Roll-Out of Smart Meters*, London: The Stationery Office (http://www.nao.org.uk/wp-content/uploads/2011/06/10121091.pdf).

National Audit Office (2013), *The CRC Energy Efficiency Scheme*. London: National Audit Office (http://www.nao.org.uk/wp-content/uploads/2012/08/ECCC_briefing_CRC_scheme.pdf).

NERA (2012), 'The Costs of the Energy Company Obligation' Report (http://www.energy-uk.org.uk/publication/finish/5/752.html).

Newbery, D. (2012), 'Progress of the UK electricity market reform' (http://www.eprg.group.cam.ac.uk/wp-content/uploads/2012/10/Progress-on-Electricity-Market-Reform-R1acc.pdf).

Office for Budget Responsibility (2013), 'Economic and Fiscal Outlook: March 2013' (http://budgetresponsibility.independent.gov.uk/pubs/Copy-of-March-2013-EFO-charts-and-tables.xls).

Ofgem (2004), 'The renewables obligation – Ofgem's first annual report' (http://www.viewsofscotland.org/library/docs/OFGEM_First_Annual_Report_Feb04.pdf).

Ofgem (2011a), 'What can behavioural economics say about GB energy consumers?' (http://www.ofgem.gov.uk/Markets/RetMkts/rmr/Documents1/Behavioural_Economics_GBenergy.pdf).

Ofgem (2011b), 'Renewable Heat Incentive guidance, Volume One: eligibility and how to apply' (http://www.ofgem.gov.uk/e-serve/RHI/howtoapply/Documents1/RHI_Guidance_Document_Vol_One.pdf).

Ofgem (2012a), 'Warm Home Discount Scheme Annual Report Scheme Year 1' (http://www.ofgem.gov.uk/Sustainability/Environment/WHDS/Documents1/WHD_AR_08_Oct_2012.pdf).

Ofgem (2012b), 'The Renewable Heat Incentive: consultation on interim cost control' (https://www.gov.uk/government/uploads/system/uploads/attachment_data/file/42906/4729-rhi-consultation-interim-cost-control.pdf).

Ofgem (2012c), 'Non-domestic Renewable Heat Incentive: a government response to "Providing Certainty, improving performance" July 2012 consultation' (https://www.gov.uk/government/uploads/system/uploads/attachment_data/file/128679/Gov_response_to_non_domestic_July_2012_consultation_-_26_02_2013.pdf).

Ofgem (2013a). 'Renewables Obligation Annual Report 2011–2012' (https://www.ofgem.gov.uk/ofgem-publications/58133/ro-annual-report-2011-12web.pdf).

Ofgem (2013b). 'Feed-In Tariff (FIT): Annual Report 2011–12' (http://www.ofgem.gov.uk/Sustainability/Environment/fits/Documents1/FITs%20Annual%20Report%202011-2012.pdf).

Ofgem (2013c), 'The final report of the Carbon Emissions Reduction Target (CERT) 2008–2012' (http://www.ofgem.gov.uk/Sustainability/Environment/EnergyEff/Documents1/CERT_FinalReport2013_300413.pdf).

Ofgem (2013d), 'The final report of the Community Energy Saving Programme (CESP) 2009–2012' (http://www.ofgem.gov.uk/Sustainability/Environment/EnergyEff/cesp/Documents1/CESP%20Final%20Report%202013_FINAL%20300413.pdf).

Ofgem (2013e), 'Information Note 13 February 2013 – the Renewables Obligation buy-out price and mutualisation ceiling 2013–4' (http://www.ofgem.gov.uk/Sustainability/Environment/RenewablObl/Documents1/buy-out%20price%20and%20mututalisation%20ceiling%202013_14.pdf).

Pearce, D. (2005), 'The United Kingdom Climate Change Levy: a study in political economy', Report for the Joint Meetings of Tax and Environment Experts under the OECD's Fiscal Affairs Committee and Environment Policy Committee, COM/ENV/EPOC/CTPA/CFA(2004)66/FINAL.

Pigou, A,C. (1920), *The Economics of Welfare*, London: Macmillan and Co.

Pizer, W. (1999), 'The optimal choice of climate change policy in the presence of uncertainty', *Resource and Energy Economics*, vol. 21, pp. 255–87.

Pollitt, M. and Shaorshadze, I. (2011), 'The role of behavioural economics in energy and climate policy', University of Cambridge Electricity Policy Research Group Working Paper no. 1130 (http://www.dspace.cam.ac.uk/bitstream/1810/242021/1/cwpe1165.pdf).

Pollitt, M. G. (2010), 'UK renewable energy policy since privatisation' EPRG Working Paper no. 1002, Electricity Policy Research Group, Cambridge University, January.

Popp, D. (2006), 'R&D subsidies and climate policy: is there a "free lunch"?', *Climatic Change*, vol. 77, pp. 311–41.

Preston, I. and Croft, D. (2012), *Environmental and Social Levies: Past, Present and Future*, Bristol: Centre for Sustainable Energy (http://www.consumerfocus.org.uk/files/2012/12/Past-and-future-trends-in-environmental-and-social-levies-ACE-CSE.pdf).

Preston, I., White, V., Browne, J., Dresner, S., Ekins, P. and Hamilton, I. (2013), *Designing Carbon Taxation to Protect Low Income Households*, York: Joseph Rowntree Foundation (http://www.jrf.org.uk/sites/files/jrf/carbon-taxation-income-full.pdf).

Quirion, P., and Demailly, D. (2008), 'European Emission Trading Scheme and competitiveness: a case study on the iron and steel industry', *Energy Policy*, vol. 30, pp. 2009–27.

Ramsey, F. (1927), 'A contribution to the theory of taxation', *Economic Journal*, vol. 37, issue 145, 47–61.

Redpoint (2012),'Modelling the trajectory of the UK power sector to 2030 under alternative assumptions - A report for the Committee on Climate Change'. London: Redpoint Energy Ltd. (http://hmccc.s3.amazonaws.com/2012%20Progress/C2Modelling%20the%20trajectory%20of%20the%20UK%20power%20sector%20v1%20%282%29.pdf).

Resch, G., Faber, T. , Haas, R , Ragwitz, M., Held, A., Konstantinaviciute, I. (2006). 'Potentials and cost for renewable electricity in Europe – the Green-X Database on dynamic cost-resource curves', Report (D4) of the OPTRES Project, Energy Economics Group, University of Vienna, February (http://www.optres.fhg.de/results/Potentials%20and%20cost%20for%20RES-E%20in%20Europe%20%28OPTRES%20-%20D4%29.pdf).

Ritz, R. A. (2009), 'Carbon leakage under incomplete environmental regulation: an industry-level approach', University of Oxford, Department of Economics, Discussion Paper Series, Paper no. 461.

Roberts, S. and Redgrove, Z. (2011), 'The Smart Metering Programme: a consumer review', Bristol: Centre for Sustainable Energy (http://www.cse.org.uk/downloads/file/smart_metering_programme_consumer_review_for_Which.pdf).

Taschini, L. (2013), *Options for Structural Measures to Improve the European Union Emissions Trading System: Response to a European Commission Consultation.* London: Grantham Research Institute (http://www.cccep.ac.uk/Publications/Policy/docs/european-union-emissions-trading-system-response-european-commission-consultation.pdf).

Thaler, R. (1999), 'Mental accounting matters', *Journal of Behavioral Decision Making*, vol 12, issue 3, pp. 183–206.

Toke, D. (2007), 'Renewable financial support systems and cost-effectiveness', *Journal of Cleaner Production*, vol. 15, issue 2007, 280–287.

Varma, A., Milnes, R., Miller, K., Williams, E., De Bruyn, S. and Brinke, L. (2012), 'Cumulative impacts of energy and climate change policies on carbon leakage' Report, CE-Delft.

Weitzman, M. L. (1974), 'Prices versus Quantities', *Review of Economic Studies*, 41, 477-91

Wilson, C. and Dowlatabadi, H. (2007), 'Models of decision-making and residential energy use', *Annual Review of Environment and Resources*, vol. 32, pp. 169–203.

Woodman, B. and Mitchell, C. (2011), 'Learning from experience? The development of the Renewables Obligation in England and Wales 2002–2010', *Energy Policy*, vol. 30, pp. 3914–21.

Appendix A. Current UK policies related to energy use

This appendix sets out the main policies that affect energy use by households and firms in the UK. We focus on policies currently in place (as of fiscal year 2013–4), though also include the Smart Meter Rollout and aspects of the Electricity Market Reform (EMR) package which are known to be coming in the near future, as well as two recent energy-efficiency related policies (CERT and CESP) which concluded at the end of 2012 but which still have some legacy effects for households at the moment as final measures are delivered under the schemes.

EU ETS Emission Trading System (EU ETS)

Aim	**The EU Emission Trading System (EU ETS) is a cap-and-trade scheme for direct emissions from energy-intensive facilities (traded sectors). The aim of the scheme is to reduce emissions in a cost effective manner, allowing companies to trade emission allowances and thereby determine how and where they reduce emissions.**
Sectors	The system covers emissions of CO_2 from power and heat generation plants, commercial aviation, and a wide range of energy-intensive industry sectors. These include oil refineries, steel works and the production of iron, aluminium, metals, cement, lime, glass, ceramics, pulp, paper, cardboard, acids and bulk organic chemicals, ammonia, and petrochemicals. From 2013 the scheme also includes nitrous oxide (N_2O) emissions from the production of certain acids and perfluorocarbons from the aluminium sector.
Eligibility	There are sector-specific thresholds determining eligibility. For instance, currently several activities (fuel combustion, metal production, etc.) are covered only if they have a net heat excess of 20 MW. The full list of categories of activities and thresholds to which the scheme applies can be found in Annex I to the EU ETS Directive (2009/29/EC)[a]. From 2013, Member States are allowed to exempt, under certain conditions, small installations, such as those with reported emissions below 25,000 tCO_2e. For commercial airlines, the system covers CO_2 emissions from flights within and between countries participating in the EU ETS (excluding Croatia until 2014).
Timeframe	The EU ETS was introduced in 2005 and has been divided into 'trading periods'. Phase 1: 2005-7; Phase 2: 2008–12; Phase 3: 2013–20; Phase 4: 2021–8.
How it works	The EU ETS works on a 'cap and trade' basis. A 'cap' or limit is set on certain GHG emissions (mostly CO_2) allowed by all participants and converted into tradable emission allowances (EUAs). These are allocated to participants via a mixture of free allocation and auctions. One allowance gives the holder the right

	to emit one tonne of CO_2 (or its equivalent, CO_2e). Participants must monitor and report their emissions each year and surrender enough emission allowances to cover their annual emissions. Those who are likely to emit more than their allocation can either take measures to reduce their emissions, or buy additional allowances either from companies who hold allowances they do not need or from auctions held by member states (Department of Energy and Climate Change, 2013g). Under Phases 1 and 2 allowances were allocated to industrial operators on the basis of national emissions caps (National Allocation Plans) agreed between the European Commission and individual member states. Allowances could be auctioned off, but the majority were allocated for free (grandfathering). In the current Phase 3, the emissions cap was set at an EU level. The cap was 2.04 billion tonnes of CO_2e in 2013, to be reduced by 1.74% each year (approximately 37 million tonnes of CO_2e). There is also a progressive shift towards the auctioning of allowances in place of free allocation. For aviation, a separate cap of 210 $MtCO_2$ per year has been agreed which remains fixed until 2020.
Revenues	The Office for Budget Responsibility (2013) estimates that revenues from UK auctions of ETS allowances will be £700 million in 2013–4, rising to £900 million in 2017–8.
Emissions	It has been estimated that by 2020 emissions from sectors covered by the EU ETS will be 21% (500 $MtCO_2e$) lower than in 2005. The separate cap on the aviation sector for the whole 2013–20 trading period is 5% below the average annual level of emissions in the years 2004–6 (European Commission 2013).
Evidence	An early analysis by Ellerman and Buchner (2008) revealed that, in the first two years of the EU ETS, CO_2 emissions were about 3% lower than the allocated allowances. This was associated to an estimated over-allocation of about 125 million allowances. There was also evidence, however, that, despite such over-allocation, the scheme led to an abatement of around 50–100 million tonnes of CO_2 per year. A more recent review by Martin, Muûls and Wagner (2012) revealed that the EU ETS may have led to abatement in the power sector, but the evidence for wider emission reductions in participating firms is not conclusive. No clear results were found for the economic performance of regulated firms, or on innovation. Work by De Bruyn, Nelissen and Koopman (2013) suggested that carbon leakage is a less important concern than initially thought. They find that, if the 2009 allocation of free permits had been based on more realistic assumptions, the sectors deemed at risk of carbon leakage would have fallen from the current 60% of sectors (representing 95% of industrial emissions) to 33% (10% of emissions).

[a] http://eur-lex.europa.eu/LexUriServ/LexUriServ.do?uri=OJ:L:2009:140:0063:0087:EN:PDF .

Carbon Price Floor (CPF)

Aim	**The Carbon Price Floor (CPF) is a minimum target price (implemented through a tax, the Carbon Price Support Rate, CPSR) on fuels used for power generation in the UK, and is designed to 'top up' the EU ETS carbon price. It is based on the carbon content of the fuels, and is aimed at providing an incentive to invest in low-carbon power generation by creating support and certainty for the carbon price in the UK's electricity generation sector.**
Sectors	Energy generation sector
Eligibility	The CPF applies to UK generators of fossil-fuel-based electricity, including combined heat and power (CHP) operators and auto-generators; those supplying such generators; and electricity utilities.
Timeframe	The CPF came into effect on 1 April 2013.
How it works	The CPF is designed to 'top up' the EU ETS carbon price at a combined level set at £16/tCO_2 in 2013, and increasing linearly each year to reach £30/tCO_2 in 2020 and £70/tCO_2 in 2030 (in 2009 prices). To bridge the government's target carbon price (the floor) and the market price for carbon in the EU ETS, special CPSR of the CCL (for gas, coal and LPG) and of the fuel duty (for oils) are imposed on the fossil fuels used for electricity generation. These rates are announced two years in advance to allow generators time to plan, and therefore rely on future estimates of ETS prices. The CPSR for 2013–4 and for 2014–5 are £4.94/tCO_2 and £9.55/tCO_2 respectively.
Revenues	The 2012 Budget estimated that the CPSR would raise £615 million in 2013–14.
Emissions	Over the period to 2030, the CPSR is expected to reduce emissions from UK electricity generation by a total of 261 MtCO_2 compared with the baseline (HM Treasury and HM Revenues and Customs, 2010a). However, analysis by the Institute for Public Policy Research (2011) shows that the CPF will not reduce carbon emissions at the European level and could have potentially detrimental effects for the EU ETS scheme by putting downward pressure on the price of carbon.[b]
Evidence	The extra cost of the policy is expected to increase the wholesale price of electricity by £8/MWh in 2015–16, from a baseline spot price of £47/MWh (Institute for Public Policy Research , 2011).
Impact on fuel poverty	The Government estimates that the introduction of the CPF is expected to increase fuel poverty by 30,000-60,000 households per year in 2013 (HM Treasury and HM Revenues and Customs, 2010b).

Climate Change Levy (CCL)

Aim	**The CCL is a tax on energy used by industry and the public sector designed to incentivise energy efficiency and emission reductions by increasing the effective price of energy.**
Sectors	Applies to industry, commerce, agriculture, public administration and other services.
Eligibility	All energy users, except domestic, charities (for non-business uses) and small activities (businesses consuming below 1,000 kWh/month of electricity and 4,397 kWh/month of gas). Other exemptions include fuels used for some forms of transport (e.g. rail and ferry); fuels covered by fuel duty; fuels used for combined heat and power (CHP) and electricity production (covered by CPF); and energy sourced directly from renewables (HM Revenue and Customs, 2012b).
Timeframe	The CCL was introduced in April 2001.
How it works	The levy is charged on electricity, gas, liquefied petroleum gas and solid fuels (such as coal), at the point of supply to business. The tax is paid by the companies supplying energy, which pass on the cost to customers as higher prices. The rates from 1 April 2013 are £5.24/MWh for electricity, £1.82/MWh for gas, £11.72/tonne for LPG and £14.29/tonne for other solid fuels (Department of Energy and Climate Change, 2013h). Energy-intensive firms can receive discounts on the CCL if they commit to energy efficiency or carbon-saving targets by joining a CCA (see below). As from April 2013 special carbon price support rates (CPSR) of CCL apply to gas, solid fuels and LPG used in electricity generation (see Carbon Price Floor). The CCL was originally designed to be revenue neutral, with revenue raised offset by cuts in employers' national insurance contributions (NIC). Mirrlees et al. (2011, Chapter 11) suggest that the cut to NICs cost rather more than the revenue raised from the CCL.
Revenues	Receipts from the CCL amounted to £636 million in 2012–3 (HM Revenue and Customs, 2013a.).
Emissions	The CCL was projected to reduce CO_2 emissions by 12.8 $MtCO_2$ (or 3.5 MtC) by 2010.[e] Between 1999 and the end of 2005, the CCL was estimated to have saved a cumulative total of around 60.5 $MtCO_2$.[e]
Evidence	According to Martin, de Preux, and Wagner (2011), the CCL had a strong negative impact on energy intensity, especially at larger and more energy-intensive plants. This effect was mainly driven by a reduction in electricity use and translated into a reduction in CO_2 emissions (see above).[f] Evidence also indicates that the CCL tax regime led to larger reductions in energy intensity and electricity use in comparison to the CCA (see below).

	Martin and Wagner (2009) also found that the CCL increased firms' innovative activity, as measured by intensity to patent. However, no evidence exists of any substantial impact on the economic performance of plants in terms of employment, gross output or productivity.

Climate Change Agreement (CCA)

Aim	**CCAs are voluntary agreements between certain energy-intensive users and the government. They allow eligible industries to claim a discount on the Climate Change Levy (CCL), provided they meet targets for improving their energy efficiency or reducing their carbon emissions.**
Sectors	CCAs can be adopted by firms in the following sectors (Department of Energy and Climate Change, 2013i): aluminium; cement; ceramics; chemicals; food and drink; foundries; glass; non-ferrous metals; paper; steel; and around 20 smaller sectors (microelectronics, distillers, textiles, supermarkets, craft baking etc.).
Eligibility	From 2006, eligible sectors are those that have an energy intensity of at least 10%. Energy intensity is the ratio of energy costs to the production value of the sector. Alternatively, sectors should have an energy intensity of 3% or more, and the industry import penetration ratio must be 50% or more. The import penetration ratio is the total value of sector imports, divided by the sum of UK sector sales and net imports (the total sales value of imports minus the total value of sector exports). This ratio is calculated for the sector as a whole to determine its exposure to international competition (Department of Energy and Climate Change, 2013i).
Timeframe	CCAs were launched in March 2001. The first scheme period, administered by DECC, came to an end in March 2013. A second period, administered by the Environment Agency, runs from April 2013 to 2023.
How it works	Companies signing up to a CCA need to set up and comply with targets to increase energy efficiency or reduce CO_2 emissions. There are two stages of a CCA. First, sector-level 'umbrella' agreements are made between DECC and the sector or trade association. These set broad commitments for the industry sectors. Second, 'Underlying Agreements' are made on an individual basis between DECC and the facility operator. These contain targets allocated by the sectors to the each operator within the same sector. Meeting any target set out under a CCA allows a facility to claim a reduction in the CCL. This reduction was 80% for all eligible fuels (electricity, gas, LPG and solid fuels like coal) until April 2011. This was reduced to 65% for all fuels, with the exception on electricity, after this date. In March 2011 the reduction for electricity was raised to 90%.

Revenues	The CCA is effectively a reduction in an environmental tax (the CCL). The revenue forgone by continuing the CCAs for industrial, commercial and agricultural sectors between 2013 and 2023 is estimated to be approximately £2.1bn (£1.55bn for industrial and commercial sectors, and £510m for agricultural sectors) (HM Revenue and Customs, 2013b.c) .
Emissions	In the fifth target period (2010) it was estimated that 28.5 $MtCO_2$ was saved each year in comparison to sectoral baselines (baseline years vary, depending upon the sector, and range from 1990 to 2008) (Department of Energy and Climate Change, 2011b).
Evidence	Early studies (AEAT 2004; Barker, Ekins and Foxon, 2007; Ekins and Etheridge 2006) have found that the CCA scheme contributed to substantial carbon savings compared with counterfactual baseline emissions. Ekins and Etheridge (2006) suggest that, even though the targets imposed by CCAs on plants were met relatively easily, the agreements contributed to increased energy savings through their 'awareness effect'. In contrast, Martin, de Preux and Wagner (2011) indicated that the targets imposed by the CCA led to lower reductions in energy consumption than the price incentive provided by the CCL. Overall, the full-rate CCL led plants to reduce growth in CO_2 emissions by between 5 and 26% more than the CCA targets. Furthermore, they show that the CCA firms experienced a decline in patenting relative to other firms. They found no evidence of substantial effects of the CCL on the economic performance of plants, and concluded that the tax discount granted to plants in a CCA cannot be justified as a means to mitigate negative impacts on economic performance.

Carbon Reduction Commitment Energy Efficiency Scheme (CRC)

Aim	**The CRC requires participants to monitor and report their energy consumption, and buy allowances equal to their CO_2 emissions. The aim of the scheme is to encourage energy efficiency across large public and private organisations that are not already covered by the EU Emission Trading System (EU ETS) and Climate Change Agreements (CCA).**
Sectors	Large public and private sector organisations, such as supermarkets, water companies, banks, local authorities (excluding state-funded schools from April 2013) and all central government departments.[a]
Eligibility	Organisations qualify as CRC participants on the basis of their electricity usage. They are eligible if they (and their subsidiaries) had at least 1 electricity meter settled on the half-hourly market during 2008, and consume more than 6,000 MWh per year.[b] The CRC targets CO_2 emissions not already covered by CCAs and the EU ETS. However, organisations that participate in the ETS and/or have a CCA can still be covered by the CRC for the emissions which are not fully regulated under those policies (for example, a single entity or a member of a group who has less than 25% of its total emissions

	covered by a CCA will still be subject to the CRC) (Environment Agency 2013).
Timeframe	The CRC started in April 2010. Following several changes, a fully revised and simplified version of the scheme was eventually introduced with the CRC Energy Efficiency Scheme Order 2013[c], which came into force in May 2013. The majority of the proposals will be introduced in 2014–5. On April 2014, five consecutive five-year phases will start; a final phase of four years will start on April 2039. A review of the scheme is expected in 2016.
How it works	Participant organisations are required to report their electricity and natural gas use on an annual basis (initially the CRC covered 29 fuels, but these were narrowed down to 2 after simplifications adopted in 2012). Natural gas is to be reported when used for heating purposes only. Reporting involves the submission of a Footprint Report during the first year of each phase, and an Annual Report at the end of each compliance year thereafter. The CRC performance league tables (comparing each participant with others in the scheme) have been replaced by a publication of participants' energy use and emissions from July 2013. From April 2013 (Phase 2), participants can buy allowances for each compliance year at two sale events (in Phase 1 a single sale of allowances took place at the beginning of the year). The first sale is based on predicted emissions and the second is a 'buy to comply' sale after the end of the compliance year. Prices are expected to be higher at the second sale (Carbon Trust, 2012). Between 2011 and 2014, the allowance price is £12 per tonne of CO_2. Firm costs (excluding administrative costs) are equal to implied CO_2 emissions (calculated through fuel use) multiplied by the allowance price (Environment Agency 2013). From 2013–4, there will be a transition from a fixed price to the auctioning of a fixed allowance total.
Revenues	Actual revenues of £700 million in 2011–12. Forecast revenues of £700 million in 2012–3, £700 million in 2013–4 (HM Treasury, 2013).
Emissions	The sectors targeted by the scheme generate over 10% of UK CO_2 emissions, or around 55 $MtCO_2$. The policy aims to reduce carbon emissions by at least 4 $MtCO_2$ per year by 2020 (Carbon Trust, 2012).
Evidence	A key issue for the CRC scheme is administrative complications and costs. The Government has proposed a new version of the CRC that is meant to simplify some of these. The DECC Impact Assessment, conducted prior to the implementation of the revised CRC, estimates that the simplified CRC scheme could involve a 55% reduction in administrative costs (a reduction of £275m in 2012 prices; see Department of Energy and Climate Change, 2013j), with negligible effects on carbon savings and energy use compared with the previous scheme (National Audit Office, 2013).

[a] https://www.gov.uk/crc-energy-efficiency-scheme .

[b] http://www.doeni.gov.uk/niea/cca exemptions 2010.pdf .

[c] http://www.legislation.gov.uk/uksi/2013/1119/pdfs/uksi 20131119 en.pdf

Emission Performance Standards (EPS)

Aim	The goal of the EPS is to limit CO2 emissions by setting mandatory emission standards for new fossil fuel power stations
Sectors	Energy sector.
Eligibility	New carbon-intensive (i.e. unabated fossil fuel) power stations with capacity at or over 50 MW.
Timeframe	This standard will come into force in 2014 and will apply until 2045.
How it works	The Emissions Performance Standard (EPS) is an annual limit on the total amount of CO2 per unit of installed capacity that new fossil fuel power stations are allowed to emit. The EPS will initially be set at a level equivalent to 450g CO2/kWh (at baseload) until 2045 for all new fossil fuel plants, except Carbon Capture and Storage (CCS) demonstration plants. The level of the EPS on the date of consent of a new power station will apply for the economic life of the installation. This process is known as grandfathering.[a] In practice, this implies that new coal plants will be able to operate only if they use CCS on at least part of their capacity. The EPS is unlikely to impact on new CCGT gas capacity, as the emissions limits are higher than the emissions factors of CCGT gas plants that are expected to be built in the future.
Revenues	Not applicable.
Emissions	No estimate available.
Evidence	The Government's impact assessment estimated that the EPS will not have an impact on investor decisions regarding new plants, relative to the estimated baseline (which includes all EMR policies except the EPS), as existing policies are expected to be able to prevent new unabated coal to come into the system even without EPS (Department of Energy and Climate Change, 2012j). The effect of the EPS will rather be to act as a 'back-stop' ensuring that any new carbon emitting generating capacity is run consistently with the decarbonisation targets, overcoming potential uncertainties in the baseline. The Energy and Climate Change Committee criticised the EPS on the ground that it 'would be pointless and would merely add complexity to an already overly complicated package of reforms' (House of Commons, 2012a). Furthermore, despite common agreement that grandfathering is necessary to ensure sufficient investment in new gas-fired generation, there has been some concern that a long grandfathering period risks locking the UK into a high-carbon growth path. The Committee on Climate Change (2012) for instance stressed that the EPS 'carries the risk that there will be too much gas-fired generation instead of low carbon investment'. An additional concern raised is that new coal-fired plants that contain some element of CCS, such as a pilot CCS system, will be allowed to continue operating without being subject to the EPS,

	even if they could still produce significant GHG emissions (House of Commons, 2012a). The Government believes the number of such projects would be limited.[a]

[a]http://www.independent.co.uk/environment/climate-change/greens-warn-of-a-return-to-era-of-dirty-coal-7791513.html.

Renewable Obligation (RO)

Aim	**The Renewable Obligation (RO) is a mechanism for incentivising large-scale renewable electricity generation in the UK by requiring electricity suppliers to source a specified proportion of the electricity they provide to customers from renewable sources**
Sectors	Energy sector.
Eligibility	To be eligible, renewable electricity projects that seek RO support should have capacity over 5 MW.
Timeframe	The RO scheme was introduced in April 2002. From March 2017 onwards all new renewables stations will be supported by the new Contract-for-Difference.
How it works	The RO scheme requires retail electricity suppliers to source a certain fraction of their annual electricity from producers using specific renewable sources. In exchange for purchasing renewable electricity, suppliers receive Renewables Obligation Certificates (ROCs). Originally one ROC was issued for each MWh of eligible renewable output generated. Since April 2009, technology 'banding' has been introduced: different levels of financial support are awarded to generators based on their generation technology, to reflect differences in technology costs and market readiness (House of Commons, 2012). For example, in 2013–4 the electricity generated from landfill gas (a more mature technology) received 0.25 ROC per MWh, while emerging technologies such as tidal steam and waves received 5 ROC per MWh. For most technologies the ROC bands are due to decrease across time[a]. Between 2003 and 2009, the level of the annual obligation was set as a series of fixed yearly targets. This was 0.03 ROC per MWh of total electricity supplied in 2003–4. This level increased annually, with a target level of 0.104 ROC/WMh in 2010–11 (Ofgem 2004). Following legislation introduced in 2009, the level of the obligation is announced by DECC six months prior to the commencement of each obligation period. The obligation is set at the highest of two levels. The first is a fixed, pre-announced target in the same style as the obligation from 2003 to 2009. The second is the total amount of renewable electricity expected to be generated during the period, with an additional 10% 'headroom'. This was meant to increase certainty to generators and help keep ROC price stable, by reducing the possibility of supply exceeding the obligation in any given year and therefore crashing the value of the ROC. However, it has arguably increased long-term uncertainty for energy suppliers, as obligation levels are now set on a yearly basis. The current obligation level for April 2013 to 31 March 2014

	is 0.206 ROCs per MWh supplied in England and Wales.[b] Suppliers who do not purchase sufficient ROCs to cover their obligation must 'buy out' the difference by paying a set price per MWh (buyout price) set each year by the government. This was set at £42.02 per ROC in 2013–4 (Ofgem, 2013e). The buyout revenue is then recycled to participating suppliers in proportion to their ROCs. Suppliers pass on the costs of RO support evenly across all electricity sales (Department of Energy and Climate Change (2011d).
Revenues	The total value of the ROCs presented for compliance in 2011–12 was £1.45 billion (Ofgem (2013a). In 2011, the government set a cap on the cost of ROs as part of their decision to introduce a cap on levy-funded spending (Levy Control Framework; see HM Treasury 2011b). The cap for 2011–2015 was set at £9.6 billion.
Emissions	During 2011–2013, the generation of renewable electricity as estimated by the redeemed ROCs amounted to a saving of 15.1 $MtCO_2$ (Ofgem, 2013a).
Evidence	Ofgem's annual review (2011–12) revealed that, since the introduction of the RO, renewable generation accounted for more than 10% of all UK supplies of electricity (although it is unclear whether this is the effect of the RO alone). The cost of saving each tonne of CO_2 under the scheme in 2011–12 was estimated at £96.61 (Ofgem, 2013a). Analysis by Mitchell, Bauknecht and Connor (2006) suggests that the RO system is less effective at increasing the share of renewables when compared with the German system of fixed feed-in tariffs. This is because it exposes RES generators to higher price, volume and balancing risk. A more recent paper by Woodman and Mitchell (2011) observed that aspects of these risks have been addressed in subsequent RO reforms, but some risks have not yet been eliminated. They stressed that, even under the redesigned RO, generators have no guaranteed market, as they are still required to negotiate a price for their output with suppliers. The resulting uncertainty about future revenue penalises small independent developers and new entrants in particular. Research by Toke (2007) on wind power development found that the German feed-in tariff scheme has been more cost-effective than the British RO. Similar conclusions were reached by Butler and Neuhoff (2008) They suggested that, once the difference in the wind resource is taken into account, the price paid for wind energy has been lower in Germany than in the UK, and that deployment has been greater in Germany, both in relation to absolute capacity installed and to policy targets.

[a]Banding levels until 2017 can be found at https://www.gov.uk/calculating-renewable-obligation-certificates-rocs.

[b]https://www.gov.uk/government/policies/increasing-the-use-of-low-carbon-technologies/supporting-pages/the-renewables-obligation-ro.

Feed-in Tariffs with Contracts for Difference (CfD FITs or CfDs)

Aim	**The CfD FITs are long-term contracts to provide developers of eligible low-carbon generation, including nuclear, renewables and CCS, with a stable revenue stream, enabling investment in low carbon energy.**
Sectors	Energy sector.
Eligibility	Renewable generation projects which are currently eligible but do not receive support under the Renewables Obligation (RO), in addition to CCS and nuclear generation.
Timeframe	The EMR is planned for implementation in 2013, with the first CfDs expected in the second half of 2014. The first capacity auction will take place by the end of 2014.
How it works	A CfD contract enables the generator to stabilise its revenues at a pre-agreed 'strike price' for the duration of its contract. Payments can flow from the contract counterparty to the generator, and vice versa (hence the 'two-way' CfD FIT). Generators receive payments from the counterparty when the electricity price received in the market (the so-called reference price) is below the agreed strike price. Generators make payments to the counterparty when the reference price exceeds the strike price. In each case the payment is the size of the difference between the two prices. The reference price attempts to reflect the wholesale electricity price, and implicitly should approximate the price the generator can expect to realise in a developed market. The two-way nature of the scheme ensures that consumers are protected from paying generators where the wholesale price of electricity would be sufficient to support the generation of renewable electricity. The Government announced draft strike prices for the period 2014–19 in June 2013. Proposed prices for 2014–5 range from £65/MWh for landfill gas to £305/MWh for wave and tidal stream; onshore and offshore wind are expected to have a strike price of £100/MWh and £155/MWh respectively (Department of Energy and Climate Change, (2013k). Final strike prices will be set in December 2013. The programme will be included in the Levy-Funded Framework.
Revenues	No estimate available.
Emissions	No estimate available.
Evidence	Department of Energy and Climate Change (2013b) modelling indicates that EMR has the potential to reduce average annual household electricity bills by between 6 to 8% (£38 to £53) over the period 2016-2030 (relative to a BAU scenario with a decarbonisation target of 100 g CO_2/kWh and using existing policy instruments; cf. Department of Energy and Climate Change, 2013l). CfD FITs have been criticised for their overly complex implementation process (House of Commons, 2012a). The scheme

	has also received criticism in relation to the proposed 'formatted' contract design, and the extent to which low-carbon technologies differ (and therefore require separate estimations and forecasts of future prices and costs; see Helm, 2012; Newbery, 2012). It has also been argued that a transition to auctioning and a competitive process of allocating contracts will prove very difficult (Helm, 2012).

Small-scale Feed-in Tariffs (FITs)

Aim	**Feed-In Tariffs (FITs) provide a long-term financial incentive for businesses and households to generate renewably sourced energy on a small scale (up to 5 MW).**
Sectors	Domestic and non-domestic sector
Eligibility	Eligible installation includes technologies that generate up to **5 MW** of energy using one of the following technologies: wind or water turbines; anaerobic digestion; solar photovoltaic (PV) panels; or micro combined heat and power.
Timeframe	The scheme was introduced on 1 April 2010.
How it works	Large energy suppliers (those with a minimum of 50,000 domestic electricity supply customers) are obligated to make FIT payments to all households and businesses with accredited installations. Payments are composed of two distinct components: 1) The 'generation tariff' is paid to all agents that produce energy, regardless of who eventually uses the energy. Set payments are made for each kWh of energy produced, and vary by the type and scale of generation, and the date on which the technology was accredited (the 'reference date'). 2) The 'export tariff' is a payment made for each kWh of energy that is exported to the national grid. Payments on a per kWh basis do not vary according to installation type or size.[a] Installations that produce more than 30 kWh have their exports measured. Installations that produce a smaller amount are typically not metered, but receive payments in line with the assumption that 50% of electricity generated is exported. Variation in the generation tariff is substantial, arising from differences in technology, capacity, and reference date. Installations registered under the scheme are guaranteed to receive payments for at least 20 years. Payments are indexed to the Retail Price Index and guaranteed for 20 years (25 for PV installations). Total FIT payments made from April 2012 to March 2013 are estimated at £497 million. The total estimated cost of the FIT scheme to 2030 and beyond is estimated at £7.9 billion (in 2009 values).

Expenditures	The scheme cost £151 million in 2011–12, considerably higher than the £94 million budgeted as part of the 2010 Spending Review. The 2010 Spending Review sets out annual budgets up to and including 2014–5; £196 million in 2012–3, £328 million in 2013–4, and £446 million in 2014–5.
Evidence	Ofgem (2013b) suggests that pre-announced changes in tariff levels significantly affected the number of PV installations registered in 2011 and 2012.[c] Huge increases in new registrations are observed prior to substantial tariff cuts in December 2011 as agents hurried to register installations before the deadline. Similar observations occur prior to further tariff changes in March, August and November 2012.

[a]Payments do vary by installation date following the Phase 2A review, with installations accredited after 1 December 2012 (or 1 August 2012 for PV technologies) receiving a greater export tariff than those with an earlier reference date.

[b]Detailed tariff rates are available at http://www.fitariffs.co.uk/eligible/levels/

Renewable Heat Incentive (RHI)

Aim	**To provide financial incentives to increase the generation of renewably sourced heat.**
Sectors	The RHI currently applies to all non-domestic sectors, including the industrial and the commercial sector, the public sector, not-for-profit organisations and producers of biomethane.
Eligibility	Eligible renewable heat technologies include solid biomass, ground and water source heat pumps, geothermal, solar thermal, biogas combustion and biomethane injection (Ofgem, 2011b). Installations must have been completed on or after 15 July 2009.
Timeframe	The non-domestic scheme (Phase I) commenced in November 2011. The domestic scheme is scheduled to begin in April 2014.[c] This was delayed from an initial intended launch date in 2012. As a forerunner to the domestic RHI scheme, the Renewable Heat Premium Payment (RHP) was introduced in 2011.
How it works	The scheme provides a per kilowatt-hour thermal (kWhth)[a] subsidy for eligible renewable heat generated from accredited installations. All subsidies are technology-specific, and are paid on a quarterly basis over a 20-year period. Subsidies also vary in value depending on the size of the installation for specific technologies. In the case of Biomass and Municipal Solid Waste technologies, a two-tier tariff is in effect to 'provide sufficient support but at the same time avoid over-subsidising' for these technologies (Department of Energy and Climate Change, 2011e). A 'Tier 1' tariff is paid for the generation of heat up to 15% of the annual load factor of the installation (equivalent to 1,314 peak load hours). Any additional generation subsequently receives a lower 'Tier 2' tariff.

	In 2013/2014 tariffs ranges were as following (dependent on size and tier): 1–8.6p for biomass boilers; 3.5–4.8p for ground source heat pumps; and 9.2p for solar thermal technologies. A proposed revision of the scheme for 2014–5 envisage higher tariffs for large biomass boilers (2p), ground source heat pumps (7.2–8.2p) and solar thermal (10–11.3p) (Department of Energy and Climate Change, 2013m)
Expenditures	The scheme has a value of £801 million between 2011–12 and 2014–5. The 2010 Spending Review set budgets of £56 million in 2011–12, £133 million in 2012–3, £251 million in 2013–4 and £424 million in 2014–5. The 2012–3 budget was reduced to £70 million in June 2012 to 'ensure the supply chain can be maintained with the available funds' (Ofgem, 2012b) This announcement also included scope for future reductions in tariffs, in order to keep expenditure within budget limits if take-up is higher than forecast levels, by introducing a 'degression mechanism' for this purpose.
Evidence	Ofgem (2012c) suggests that take-up of the policy is below that made possible by the pre-planned budgets. Even after the 2012–3 budget reduction is taken into account, expected expenditure on the scheme is expected to be only 60% of the annual budget.

[a] One kWhth is the amount of heat energy given off by a 100% efficient 1 kW electric heater left on for an hour.

Capacity Mechanism

Aim	**A capacity mechanism would make payments to generators for the availability of capacity (rather than for the electricity they produce) to provide sufficient incentives for investment in new capacity.[a] The aim is to provide an insurance policy to reduce the likelihood of future blackouts and to ensure a reliable electricity supply to consumers.**
Sectors	Energy sector.
Eligibility	The Government plans to exclude plants receiving CfD FITs from the Capacity Market to avoid overcompensation of low carbon plants.b
Timeframe	The first capacity auction could take place as early as Autumn 2014, for a delivery year of winter 2018–9.
How it works	The capacity mechanism provides payments in an attempt to ensure that the desired optimal capacity margin is met. The proposed capacity mechanism would work as follows. Following a four-year forecast of future peak demand, the total amount of capacity needed to ensure security of supply will be contracted through a competitive central auction. Providers of capacity who are successful in the auction will enter into capacity agreements, committing to provide electricity when needed in the delivery year(s) in return for a steady capacity payment. If they fail to

	provide this capacity they face penalties. The costs of the capacity payments will be shared between electricity suppliers in the delivery year (Department of Energy and Climate Change, 2012k). The first capacity market will be launched in 2014, with the aim of ensuring new capacity is in place by winter 2018.
Revenues	Not applicable.
Emissions	No estimate available.
Evidence	The Government's Impact Assessment estimated that an Administrative Capacity Market could lead to an increase in bills of around £16/year per domestic household (Department of Energy and Climate Change, 2012k). Helm (2012) argues that the overlap between CfD FITs and the capacity mechanism may complicate investment and progress of new technologies (as they both serve to secure new capacity). He therefore supports the merging of CfD FITs and the capacity mechanism into a single quantity mechanism. The Committee on Energy and Climate Change argued that the capacity mechanism should be accompanied by an enduring reliability target (i.e. a minimum level of capacity is needed). The Committee also raised concerns over the fact the proposal was based upon out-of- date assumptions and an insufficient analysis of the future risk of reliability. Further investigation on costs, and on the impact on gas generation and storage was also recommended (House of Commons (2012a).

Carbon Emissions Reduction Target (CERT)

Aim	**CERT aimed to reduce carbon emissions from the UK residential sector by placing an obligation on large energy suppliers to improve energy efficiency in the existing housing stock.**
Sectors	Domestic. Large energy suppliers were mandated to provide the service.
Eligibility	At least 40% of all actions were to be achieved in the 'priority group'. This included households with an individual aged 70 years or more, or households in receipt of a number of means-tested benefits. Savings of at least 16.2 $MtCO_2$ must be achieved in a particular subset of this group; the 'super priority group'.[a] CERT placed obligations on energy suppliers with more than 250,000 domestic customers (i.e. the six main UK energy suppliers).
Timeframe	1 April 2008 to 31 December 2012 (when CERT was superseded by ECO).
How it works	The obligation required eligible suppliers to install energy efficiency measures in residential buildings in order to create carbon savings of 154 $MtCO_2$ between 1 April 2008 and 31 March 2011. This target was later increased to 293 $MtCO_2$ over an extended period of time (up to 31 December 2012).

	Qualifying measures were split into four eligible components: 1) 'Standard actions' which achieve a reduction in carbon emissions. 2) 'Demonstration actions' which could reasonably be expected to achieve a reduction in carbon emissions. 3) 'Market transformation actions' which included the installation of solid wall insulation or micro-generation technology (and that resulted in carbon emissions reductions). 4) 'Priority group flexibility actions' which provided solid wall insulation to a priority group householder. A number of restrictions were applied in the way savings could be realised. These included: 1) An 'insulation target' required savings of 73.4 $MtCO_2$ to be realised from insulation measures installed professionally after 1 August 2010. 2) No more than 6% of reductions may be realised from demonstration and market transformation actions. 3) No more than 12.5% of priority group actions may be classified as 'flexibility actions'.
Expenditures	DECC impact assessments, conducted prior to the implementation of the policy, estimated that the delivery of CERT would cost around £3.9 billion (2010–11 prices according to Hansard[b]).
Evidence	Over the course of the policy, a total of 296.9 $MtCO_2$ of carbon savings was realised (equivalent to 101.3% of the overall CERT target). 41.3% of savings were achieved in the priority group. 16.6 $MtCO_2$ of savings were made in the Super priority and 75.1 $MtCO_2$ of savings were made toward the insulation target. Despite producing the required level of savings, no specific target was achieved by all firms. As a result, each target was considered as missed (Ofgem, 2013c).

[a]Detailed eligibility criteria for each group are available at:
http://www.legislation.gov.uk/uksi/2008/188/contents/made and
http://www.legislation.gov.uk/ukdsi/2010/9780111500095/article/3.
[b]http://www.publications.parliament.uk/pa/ld201011/ldhansrd/text/111025w0001.htm#11102553000331.

Community Energy Saving Programme (CESP)

Aim	**CESP aimed to work alongside CERT in order to reduce carbon emissions from the UK residential stock, and focused in particular on achieving these savings in areas of low income.**
Sectors	Domestic (private and social housing stock in deprived areas). Large energy suppliers and generators were mandated to provide the service.
Eligibility	Qualifying actions must be provided to domestic energy users in areas of low income. These are defined as the areas with the 10% of highest Index of Multiple Deprivation (IMD) scores in England, and the 15% of highest scores in Scotland and Wales. Actions that are counted as qualifying measures under CERT cannot be counted towards achieving the CESP savings target. CESP placed obligations on energy suppliers with more than 50,000 domestic customers and electricity generators that produce at least 10TWh per year (i.e. the six main UK energy suppliers and four generators).
Timeframe	1 October 2009 to 31 December 2012 (when CESP was superseded by ECO).
How it works	Eligible energy suppliers and electricity generators installed efficiency measures to produce savings of 19.25 $MtCO_2$ in areas of low income. Qualifying actions included solid wall insulation and the replacement of inefficient boilers.[a] A number of restrictions were placed on how these savings are achieved. A maximum 4% of savings could be achieved through loft insulation, 4% through cavity wall insulation and 1% through the provision of a home energy advice package. Obligated firms received a bonus towards achieving their target when providing actions which directly benefit more than 25% of the people in a single qualifying area. A bonus was also available for providing two or more measures to a single dwelling.
Expenditures	The cost of the scheme was believed to be £350 million at the time of the policy announcement (House of Commons, 2012c).
Evidence	Obligated suppliers achieved 16.31 $MtCO_2$ of savings by 31 December 2012. This represented approximately 85% of their overall target. Suppliers met 92% of their targets, while generators met only 36%. (Ofgem 2013d).

[a]A full list of qualifying measures is available at http://www.legislation.gov.uk/uksi/2009/1905/schedule/2/made

Energy Company Obligation (ECO)

Aim	**ECO aims to improve the energy efficiency of the domestic housing stock through the provision of insulation and efficiency measures for poor households or those in hard-to-treat properties.**
Sectors	Only the domestic sector is covered by ECO. The service is provided by large energy suppliers with the costs recouped from domestic energy users.
Eligibility	Eligibility for the affordable warmth group (see below) is based on receipt of means-tested benefits or tax credits alongside other criteria including old age, having dependent children or disability.[a] Eligibility for the carbon saving community obligation (see below) depends on living in the most deprived 15% of lower-level super output areas in Britain.[b] ECO is delivered by gas and electricity suppliers that have 250,000 domestic customers or more, and supply more than 400 GWh of electricity or 2,000 GWh of gas to domestic customers (i.e. the six main UK energy suppliers).
Timeframe	The initial phase of ECO started in October 2012. Requirements are currently set to March 2015 though it is expected to continue beyond then.
How it works	ECO effectively replaces the earlier CERT and CESP schemes (see below) and runs alongside the Green Deal. ECO is divided into three phases. Phase one ran from October 2012 to March 2013, phase two runs from April 2013 to March 2014, and phase three from April 2014 to March 2015. There are three parts to ECO with specific requirements under each part in each phase: 1. The **home heating cost reduction obligation** requires investment in measures which will achieve expected energy bill savings of £0.8 billion in phase 1 and £1.7 billion in each of phases 2 and 3. These savings are targeted on the **affordable warmth group**. 2. The **carbon saving community obligation** requires investment in measures which will achieve an expected emissions reduction of 1.4m tonnes of CO_2 in phase 1, and 2.7 million tonnes in each of phases 2 and 3. At least 15% of savings must be delivered to people in rural areas (who are also in the affordable warmth group) or rural areas adjoining the eligible regions. 3. The **carbon emissions reduction obligation** requires measures to be installed that generate expected emissions reduction totalling 4.2m tonnes of CO_2 in phase 1, and 8.4 million tonnes in each of phases 2 and 3. This includes measures which do not meet the golden rule for Green Deal (such as solid wall insulation or 'hard-to-treat' cavity walls). There are no specific eligibility criteria for this part of ECO. Installation of measures generates 'ECO points'. Suppliers are

	obliged to provide points under each part of the scheme according to the size of their customer base. ECO points can be generated by independent companies (who can sell them to energy suppliers through a brokerage market) as well as by energy suppliers themselves.
Expenditures	Department of Energy and Climate Change (2012g, Table 16) estimate the costs of ECO to energy suppliers at around £1.3 billion per year. NERA (2012) suggest the costs could be in the order of £1.7 to £2.4 billion.
Impact on efficiency	It is estimated (Department of Energy and Climate Change, 2012g) that between 2013 and 2022, 2.7 million cavity walls will be insulated, 1.64m lofts insulated or topped up with additional insulation and 0.96m solid walls insulated, though these figures include business as usual installations.

[a] See http://www.legislation.gov.uk/ukdsi/2012/9780111525456/pdfs/ukdsi_9780111525456_en.pdf.
[b]A full list of qualifying areas is given in https://www.gov.uk/government/uploads/system/uploads/attachment_data/file/48405/5536-carbon-saving-community-obligation-rural-and-low-.pdf.

Green Deal

Aim	**The Green Deal aims to improve energy efficiency of the UK building stock. It offers efficiency and insulation improvements at no upfront charge, with the cost met through loans repaid through electricity bills.**
Sectors	Domestic and non-domestic sectors are covered by the Green Deal.
Eligibility	There are no eligibility criteria on who can use the Green Deal. A total of 44 different measures are eligible for installation under the Green Deal[b]. The Green Deal cash-back incentive (see below) is only available in England and Wales.
Timeframe	The Green Deal was launched in January 2013 when the first assessments took place, though as of June 2013 no loans were actually available. Until March 2014, households will be eligible to receive cash back for measures installed under the Green Deal.
How it works	People wishing to take up the Green Deal are first visited by a registered Advisor who will assess which measures are suitable for installation and estimate the energy savings. The consumer can take the results of the assessment to a Green Deal Provider who will quote for the work and a plan for repayment (interest rate and duration) through the electricity bill. Those who take up measures early may be eligible for a Green Deal Cashback incentive which offers cash payments for different measures. If the consumer moves home then the obligation to repay falls on the next owner or tenant. The 'golden rule' of the Green Deal states that the amount repaid through the bill in the first year must be no more than

	the expected bill savings that would be made through the measures installed.[a] Bill payments can then rise by no more than 2% per year. The repayment period must also be no longer than the expected life of the measure. The golden rule does not guarantee that all households will pay lower bills following a Green Deal installation, since the actual savings will depend on household's individual energy use whereas the golden rule is based on expected savings.
Expenditures	Up to £125 million has been allocated for the early Green Deal cash back incentive. It is expected that the cumulative total loaned under the Green Deal will be £3.2– £4.1 billion by 2022 (Department of Energy and Climate Change, 2012g).
Evidence	In the first four months of the scheme, 18,816 domestic assessments were carried out. As of April 2013, there were 152 Green Deal assessor organisations and 55 providers (Department of Energy and Climate Change, 2013n). There was no formal trial of the Green Deal. A precursor 'Pay as you Save' scheme was trialled in five areas with upfront financing of up to £20,000 per household available and variable repayment durations and other incentives built in (DECC and EST, 2011). Unlike the Green Deal all loans were interest-free. Despite this relatively more generous offer than the Green Deal, only just over half of households who had an assessment went on to have measures installed.
Impact on emissions	Department of Energy and Climate Change (2012g) estimate that by 2020, Green Deal and ECO combined will reduce CO_2 emissions in the non-traded sector by 1.8m tonnes per year and reduce energy consumption by 17.1TWh.
Impact on fuel poverty	Department of Energy and Climate Change (2012g) estimate that Green Deal and ECO combined are expected to reduce fuel poverty by 125,000 to 250,000 households from 2023 (the point at which ECO costs are no longer recouped through energy bills).

[a]https://www.gov.uk/government/uploads/system/uploads/attachment_data/file/70098/Golden_Rule_29_1.pdf

[b]A list is available at http://www.legislation.gov.uk/uksi/2012/2105/schedule/made).

Products Policy

Aim	**The 'Products Policy' is an umbrella term for policies expected to affect the efficiency of energy-using and energy-related products. A significant part of the policy (and the focus of this summary) is the UK implementation of EU 'Ecodesign' Directives aimed at improving the efficiency of electrical appliances and other energy-related products through minimum energy standards. The wider Products Policy also covers mandatory energy labelling of products, supplier obligations to encourage energy efficiency, building regulations, etc.**

Sectors	Domestic and non-domestic use of products and appliances will be affected.
Coverage	The Ecodesign Directives apply to different product groups agreed at the European level. As of 2013, 12 groups have specific targets established under the Ecodesign Directive (set-top boxes, domestic lighting, non-domestic lighting, external power supplies, televisions, electric motors, heating circulators, domestic fridges and freezers, domestic dishwashers, domestic washing machines, electric fans and air conditioners). Measures have also been adopted for standby/'off-mode' energy use across all electrical appliances.
Timeframe	The Ecodesign Directive was initially implemented in 2005 and introduced to UK law through Statutory Instrument in 2007. The Directive was extended in 2009 to cover products that do not use electricity directly but affect the amount used (e.g. windows, shower heads); this was introduced to UK law in 2010. The first specific regulations were adopted in December 2008.
How it works	The Ecodesign Directive establishes a framework; regulations for specific product groups are established within this framework under 'implementing measures'. These measures are worked towards following preparatory studies, consultation and impact assessment, voted on by representatives of member states and scrutinised by the Parliament and the European Council. The implementing measures for televisions, for example, set out minimum standards for the power consumption of sets when in use, in standby and turned off (Commission Regulation 642/2009). These standards become progressively more stringent. Products which fail to meet the standards cannot be offered for sale in the EU. Product groups to be the focus of attention for future implementing measures are established under three-year Working Plans. The current period (2012–4) products include windows, power cables and computer servers (European Commission, 2012a). Products considered should have sales of at least 200,000 units across the EU and be thought to have a significant environmental impact or scope for efficiency savings.
Evidence	There is no formal evaluation of the Ecodesign Directive owing to a relative lack of data since implementing measures for specific product groups were adopted. CSES and Oxford Research (2012) suggest that for domestic appliances (televisions, fridges, dishwashers etc.) there was no clear evidence that improvements in efficiency could be attributed to the implementing measures. For lighting (banning incandescent bulbs), stand-by and heat circulators there was more evidence that the measures had improved efficiency (European Commission, 2012a).
Impact on energy use and emissions	European Commission (2012b) estimate that the Ecodesign Directive will save 366 TWh of electricity per year by 2020 across the EU, around 12% of EU-wide consumption levels in 2009 (Centre for Strategy and Evaluation Services and Oxford Research, 2012). Department for Environment, Food and Rural Affairs (2009) estimates that the wider Products Policy in the UK will save around

	14.3 million tonnes of CO_2 in 2020 relative to business as usual European Commission (2012b) . Our calculations based on the policy-specific breakdown suggest that around 9.6 million tonnes are attributed to the Ecodesign measures and 4.7 million tonnes to other aspects of the wider Products Policy.

Smart Meter Rollout

Aim	**To provide 'smart' energy meters which help energy consumers reduce energy demand and bills through real-time, visible information on energy use and encourage tariff-switching. In the longer term this could also help better balancing of energy supply and demand and tailoring of tariffs to individual energy consumption patterns.**
Sectors	Domestic and small and medium non-domestic energy users will be part of the rollout process. Large non-domestic users already have close to real-time meters. Around 30 million properties are covered, comprising 53 million gas and electricity meters.
Timeframe	The rollout is expected to happen between 2015 and 2020 (Department of Energy and Climate Change, 2013o).
How it works	Smart meters record information on energy use which can be transmitted to energy suppliers without the need for a meter reader to visit. Information on energy use can be shown to customers by means of an in-home display (IHD); the government published initial minimum technical standards that the IHD must comply with in 2012 (Department of Energy and Climate Change, 2012l). Government, industry and consumer groups are currently working together in a 'Foundation Stage' which began in 2011 to prepare the way for the mass roll-out of smart meters. The intention in this stage is that common technical standards are agreed (to ensure that meters are compatible across energy companies), that a company is appointed to manage data services and a regulatory framework agreed. The roll-out phase from 2015 will be delivered through energy companies who will visit homes and install smart meters and IHDs.
Costs and benefits	The most recent impact assessment (Department of Energy and Climate Change, 2013p) for the domestic sector suggests NPV costs between 2012 and 2030 of £12.1 billion. The largest cost is the installation and operation of the the meters themselves (£7.0 billion) The NPV benefits are £18.8 billion, including £6.3 billion from reduced energy demand and £3.4 billion saved by energy companies who no longer have to visit properties to read meters. The effect of the policy on energy bills is expected to be positive in the short-term (£7 on an average household dual fuel bill in 2015) as suppliers recoup the costs of installation through higher bills, but negative in the long-term (–£25 on a bill in 2020, –£40 in 2030) as this cost is removed and the benefits of reduced energy consumption are realised.

	Estimates for the non-domestic sector (Department of Energy and Climate Change, 2012m) are costs of £0.6 billion (£0.4 billion from meter and installation costs) and benefits of £2.9 billion (£1.7 billion from reduced energy use).
Evidence	National Audit Office (2011) note uncertainty in how consumers respond to smart meters, in particular whether changes in consumption are sustained over a long period and vary across different demographic groups. AECOM (2011) report on large-scale trials of meters in the late 2000s. They find mixed evidence on whether consumption effects vary across groups, but more compelling evidence that the information and advice supplied alongside smart meters matters for the consumption response. Roberts and Redgrove (2011) argue that delivering the roll-out through energy companies limits the ability to roll out installation on an area-by-area basis (since neighbours will have different providers), which may have been a more cost-effective approach.
Impact on energy use and emissions	**Domestic sector** (Department of Energy and Climate Change, 2013p): electricity demand is estimated to fall by 2.8% per household and gas demand by 2% (for those not using prepay) or 0.5% (for those using prepay). The estimated effect on emissions is 15.9 million tonnes of CO_2 saved from reduced gas use over the period 2012–2030. Domestic emissions from electricity use will fall by 14.5 million tonnes, but since electricity use is covered by the EU ETS there will be no net global impact on emissions from reduced electricity consumption. **Non-domestic sector** (Department of Energy and Climate Change, 2012m): electricity demand is estimated to fall by 2.8% per meter, and gas demand by 4.5%. Reduced gas use will reduce CO_2 emissions by an estimated 10.1 million tonnes. Domestic emissions from reduced electricity use will fall by 3.8 million tonnes, though again this is not a net global emissions reduction.

Warm Home Discount Scheme (WHD)

Aim	**The WHD aims to reduce the incidence of fuel poverty by giving electricity bill rebates to low income and vulnerable households.**
Sectors	Energy companies with at least 250,000 domestic customers are obligated to take part in the provision of the WHD. Domestic customers are recipients of the discount.
Eligibility	Two distinct groups are eligible for the WHD rebate: 1. A **core group** of households are guaranteed eligibility for a rebate, and consist of low income pensioners. There are two sub-groups: a. Customers aged under 75 (65 from 2014–5) who receive just the guarantee credit element of the pension credit, but not the savings credit element (for a single pensioner this amounts to a weekly income below £115.30); b. Customers aged 75 (65 from 2014–5) and over who receive the guarantee credit element of the Pension credit

	are eligible irrespective of whether they also receive the savings credit element (for a single pensioner this amounts to a weekly income below £145.40). 2. A **broader group** of other customers, with the criteria at the discretion of individual energy companies (subject to Ofgem approval). In practice, eligibility is largely determined by receipt of an income-related benefit together with having young children, older people or disabled people in the household; or by receipt of the pension credit for those not already part of the core group. Core group members automatically receive WHD, broader group members must apply.
Timeframe	The WHD was launched in 2011–12, and will run until at least **2015–6**.
How it works	The total budget is divided between core and non-core expenditures. The core component is the total cost of rebates, worth £135 each to core group members. Non-core expenditure falls into one of three activities: 1) Bill rebates (of £135) for members of the broader group. 2) Legacy spending: Under previous voluntary agreements, energy suppliers provided some customers with reduced prices ('social tariffs'). Spending on these tariffs is allowed to continue under the WHD, but gradually phased out over time. 3) Industry Initiatives: Companies are allowed to count the cost of six broad activities towards their non-core WHD expenditure: assistance with claiming benefits, referrals to energy suppliers for rebates, energy-efficiency advice, energy-efficiency training, energy debt assistance and installation of energy-efficiency measures. The core and non-core budgets are non-substitutable. Within the non-core budget, a minimum amount must be spent on rebates, and a maximum amount can be spent (individually and combined) on legacy spending and industry initiatives.
Expenditures	The WHD scheme is contained within the control framework for DECC levy-funded spending. The cost of the scheme was set at £250 million in 2011–12, £275 million in 2012–3, £300 million in 2013–4, £310 million in 2014–5 and £320 million in **2015–6**.
Evidence	Ofgem (2012a) report that £238 million was spent on the WHD in 2011–12, an underspend of £12 million against the budget.[a] This occurred due to lower than expected spending on the core group (£84 million, against a budgeted £97 million). Of an estimated 800,000 eligible recipients in the core group, around 700,000 received a payment suggesting some imperfections in the automatic payment mechanism.
Impact on fuel poverty	Department of Energy and Climate Change (2011a) estimated that the scheme would reduce fuel poverty by around 58,000 households in 2014–5.

Winter Fuel Payment (WFP)

Aim	**The WFP is aimed at supporting energy costs for older people through annual tax-free payments made in winter months.**
Sectors	Domestic
Eligibility	All UK residents over the qualifying age for pension credit on the third Monday of September are eligible for the payment. Households containing members aged 80 years or above are eligible for larger payments.
Timeframe	WFPs were first introduced in 1997/98. Additional payments were initially made to households receiving means-tested benefits. This condition was dropped in 1999/2000. Larger payments for older individuals were introduced in 2003–4.
How it works	Cash transfers are made to all eligible individuals. Individuals aged between the pension credit age and 79 years old are entitled to a payment of £200. Individuals aged 80 years or above are eligible for payments of £300. A single household may receive a maximum of £300 from WFPs in a single year. Payments are adjusted for individuals in cases where two eligible individuals live in the same household. The entire payment is made to one individual in the case where a claimant is in receipt of pension credit, income-related ESA or income-based JSA.
Expenditures	12.7 million Winter Fuel Payments were made in 2012–3 at a cost of £2.15 billion.
Evidence	Beatty et al. (2011) estimate that 41% of the WFP is spent on purchasing fuel, compared with an estimated 3% from a simple cash transfer. This suggests that a 'labelling effect' is associated with the WFP. This is in contrast to standard economic theory, which implies that the name attached to an income transfer should have no bearing on how the money is spent.

Cold Weather Payment (CWP)

Aim	**CWPs aim to support vulnerable households in meeting unexpected increases in energy costs, by providing cash transfers following periods of extremely cold weather.**
Sectors	Domestic
Eligibility	Households are eligible to receive CWP if they fulfil the following criteria: a. Receive pension credit; or, b. Receive income support, income-based jobseeker allowance (JSA) or Income-related employment and

	support allowances (ESA), and fulfil at least one of the additional criteria below: i. Receive a disability or pensioner premium ii. Receive a severe or enhanced disability premium (ESA only) iii. Receive Child Tax Credit that includes a disability or severe disability element iv. The household includes a child who is disabled v. The household includes a child aged under five years vi. Receive the support or work-related component of ESA
Timeframe	Cold weather payments were initially introduced in 1986/7.
How it works	Each UK postcode is linked to one of 92 weather centres. Cold weather periods are recorded for each centre (and their relevant postcodes) following a period of seven consecutive days between 1 November and 31 March when temperatures are recorded or forecast to be 0°C or below. A transfer is automatically paid to eligible households following each independent cold weather period. In 2012–3, payments were £25 per period.
Expenditures	In 2011–12, 5.2 million payments were made at a cost of £129.2 million. It should be noted that the cost of the scheme varies significantly on an annual basis due to changes in the severity of the weather. In the previous year, which saw a much colder winter, 17.2 million payments were made at a cost of £430.8 million.
Evidence	Beatty, Blow and Crossley. (2011) find evidence that the poorest, elderly households experience a 'heat-or-eat' trade-off in the most extreme cold weather shocks.[a] This suggests that the current system of cold weather payments (and winter fuel payment) does not fully protect the most vulnerable households.

Value Added Tax (VAT) on energy use

Aim	**VAT is the main consumption tax in the UK and is largely designed to raise revenue. The standard rate is 20% but domestic energy pays a reduced rate of 5%.**
Sectors	Domestic and non-domestic energy use.
Timeframe	VAT was introduced in the UK in 1973. Firms pay the full rate of VAT on their energy use. Domestic energy use was originally zero-rated, but VAT at 8% was introduced in April 1994 with plans to raise it to the full rate (then 17.5%) in 1995. This was not implemented and the rate was reduced to 5%, the lowest permitted under EU law, in 1997.
How it works	VAT is a proportional tax on sales at each point in the production process, with the VAT already paid on inputs

	reclaimed such that the tax is only on the 'value added' at each stage. EU Directives limit the standard rate of VAT to be at least 15%, with no more than two different reduced rates (each of at least 5%) that can apply to a restricted set of goods and services. Items not subject to VAT prior to the introduction of the EU Single Market in 1992 are permitted to remain zero-rated.
Expenditures	The cost of taxing domestic energy at 5% rather than the standard rate of 20% is estimated at around £5.2 billion in 2012–3 (HM Revenue and Customs, 2012a).

Appendix B. Calculation of implicit carbon prices

This appendix details the methodology and assumptions used to estimate the implicit marginal carbon prices imposed across fuels and end users from various policies. The findings are presented and discussed in Chapter 6.

Our main results in Chapter 6 are for 2013–4, but we also show figures for 2020 and detail here how those future values are derived. We maintain throughout an implicit assumption that policies are ultimately incident on end-user fuel prices no matter where they are formally incident in the energy supply chain.

We estimate carbon prices for four groups of end users:

- **Households**
- **Small businesses** are assumed not to face the CRC for their energy and gas use.
- **Medium businesses** are large enough to face the CRC but do not qualify for a Climate Change Agreement (CCA).
- **Large non-energy-intensive businesses** are large enough to participate in the EU Emissions Trading Scheme (ETS) directly, and so do not face the CRC, and qualify for a CCA.

Table B.1 Conversion factors

Fuel	Units	Factor	Notes and source(s)
Electricity (marginal, 2013)	tCO_2e/MWh	0.368	DECC and HMT (2012) Table 1 'Domestic' estimate of carbon content of long-run marginal electricity in 2013.
Electricity (marginal, 2020)	tCO_2e/MWh	0.293	DECC and HMT (2012) Table 1 'Domestic' estimate of carbon content of long-run marginal electricity in 2020.
Gas	tCO_2e/MWh	0.185	DEFRA and DECC (2012) Table 1C.
Coal (domestic)	tCO_2e/t	2.816	DEFRA and DECC (2012) Table 1A.
Coal (industrial)	tCO_2e/t	2.184	DEFRA and DECC (2012) Table 1A.
LPG	tCO_2e/t	2.933	DEFRA and DECC (2012) Table 1B, emissions per litre (1.53 kg of CO_2e), converted using density figure in Table 11 (1 litre = 0.522 kg).

Implicit carbon prices are calculated for four fuels:

- **Gas** (supplied as gas for heating etc.)
- **Electricity** (calculations are based on the emissions of the marginal plant, which we take to be Combined Cycle Gas Turbine, see details below)
- **Coal**
- **LPG**

We calculate the combined effect of different policies on the marginal price paid for these fuels for each end user. To convert that into a carbon price, we need to know the carbon content of a marginal unit of each fuel. We use conversion factors between fuels and carbon taken from various sources which are detailed in Table B.1. As described in Chapter 6, we assume that at the margin electricity is supplied as gas (combined cycle gas turbine (CCGT) plants) and so it is the carbon content of gas-fired electricity that matters for our analysis.

To see how these are used, consider a policy which imposed a tax of £1 per MWh on both gas and electricity use. This is equivalent to a carbon tax of £2.72/tCO_2e on electricity (£1 ÷ 0.368 tCO_2e/MWh) and £5.41/tCO_2e (£1 ÷ 0.185 tCO_2e/MWh) on gas. Since a marginal increase in electricity consumption generates more carbon than a marginal increase in gas consumption, imposing the same tax rate on each fuel must equate to a higher implicit carbon tax on gas.

At the end of this appendix, we present some results using the grid-average emissions factor for electricity instead. This answers a slightly different question (what is the effective tax rate imposed by current policies on electricity used by different groups?) to the question we are most interested in answering in Chapter 6 (to what extent do current policies give similar incentives at the margin to reduce carbon emissions by reducing fuel use for different end-user groups?) though the former question may well be of policy interest as well.

We report all prices (whether for current or future carbon prices) in 2013 terms. Where policies are priced in an alternative base year, we adjust these prices using outturn and forecast RPI inflation rates from Table 1.5 in Office for Budget Responsibility (2013), except for the estimate of the domestic VAT subsidy in 2013–4 where energy-specific indices are used (see below).

B.1 Policies included in the calculations

Here we outline the policies included in our analysis, including the rates and any other information necessary to calculate the implicit carbon price imposed.

EU Emissions Trading Scheme (ETS)

The ETS imposes a carbon price on electricity generation and on other fuels (not including gas) used as inputs by large industries.

As described above we assume the impact on final prices for electricity users is determined by the marginal generating plant (CCGT) and is fully passed through.

The ETS carbon price is determined by trading and so varies over time. We use the average clearing price for carbon permits in UK auctions held between 16 January and 19 June 2013.[74] This was €4.13 per tonne of CO_2. This is converted at an exchange rate of £1 = €1.18 taken as an approximate average of rates

[74] https://www.theice.com/marketdata/reports/ReportCenter.shtml#report/148.

observed in 2013 to mid-June 2013.[75] This gives a carbon price of **£3.50/tCO_2e.** We apply that carbon tax rate directly as applicable by fuel and user.

The 2020 ETS carbon price of **£8.82/tCO_2e** was taken from the central estimate in Table 1 of Department of Energy and Climate Change (2012n).

Carbon Price Support Rate (CPSR)

Budget 2011 set CPSR rates on fossil fuels used to generate electricity. For gas, which we assume to be the marginal generator, the rate is £0.91/MWh in 2013–4. Taking the carbon content of gas from Table B.1, this equates to a carbon tax of **£4.92/tCO_2e** on electricity use for all end users.[76]

Section 8.1 of HM Revenue and Customs (2012a) states that the Carbon Price Floor in 2020 will be £30/tCO_2. The CPSR is therefore calculated as the difference between this CPF price and the ETS price above, giving a CPSR of **£21.18/tCO_2e** on electricity use for all end users in 2020.

Climate Change Levy (CCL)

The CCL imposes a tax on end-use of fuels by businesses. Rates for 2013–4 are given in Table B.2 along with the equivalent carbon taxes calculated using the conversion factors in Table B.1.

Table B.2 Rates of CCL and implicit carbon prices

Fuel	Tax rate	Unit	Carbon tax (£/tCO_2e)
Electricity	5.24	£/MWh	**14.24**
Gas	1.82	£/MWh	**9.85**
LPG	11.72	£/tonne	**4.00**
Coal	14.29	£/tonne	**6.54**

Source for tax rates: HMRC (http://customs.hmrc.gov.uk/channelsPortalWebApp/channelsPortalWebApp.portal?_nfpb=true&_pageLabel=pageExcise_RatesCodesTools&propertyType=document&id=HMCE_PROD1_031183).

The tax rates are assumed to remain constant in real terms until 2020. However, the electricity carbon tax per tCO_2e rises because of the fall in the carbon content of marginal electricity between 2013 and 2020. The 2020 electricity carbon tax is estimated to be **£17.90/tCO_2e**.

Climate Change Agreement (CCA)

Large energy-intensive firms in industries with CCAs receive a 90% discount on the rates of CCL for electricity, and a 65% discount for other energy sources in 2013–4. These discounts are assumed to remain in place to 2020.

[75] http://fx-rate.net/GBP/EUR/.

[76] Note this is very slightly different from the 'official' policy rate of £4.94. This is due to slightly different conversion factors. We use the policy conversion factor to convert into £/MWh, and then convert back as with all other policies at a common conversion rate for tCO_2/MWh.

Rates and carbon prices for firms with a CCA are shown in Table B.3. Note that we do not attempt to calculate the implicit carbon costs imposed on different industries of the agreements they sign in order to obtain the CCA. Intuitively, the expected cost of meeting the agreement must be less than the tax savings made by paying the lower CCA rate.

Table B.3 Rates of CCA (discounted CCL) and implicit carbon prices

Fuel	Tax rate	Unit	Carbon tax (£/tCO_2e)
Electricity (2013)	0.52	£/MWh	**1.42**
Electricity (2020)	0.52	£/MWh	**1.79**
Gas	0.64	£/MWh	**3.45**
LPG	4.10	£/tonne	**1.40**
Coal	5.00	£/tonne	**2.29**

Source for tax rates: HMRC (http://customs.hmrc.gov.uk/channelsPortalWebApp/channelsPortalWebApp.portal?_nfpb=true&_pageLabel=pageExcise_RatesCodesTools&propertyType=document&id=HMCE_PROD1_031183)

Renewables Obligation (RO)

Electricity suppliers are mandated to source 20.6% of electricity from renewable sources in 2013–4.[77] The RO buyout price is set at £42.02 per MWh. Assuming that the buyout price is the marginal price, an implicit RO electricity tax rate is calculated as 20.6% of the buyout price, £8.656/MWh. Using the conversion factor for the marginal unit of electricity from Table B.1, this gives an implicit electricity carbon tax of **£23.52/tCO_2e**. We apply this to all end users under the maintained assumption of full pass-through.

For 2020 we use figures from Department of Energy and Climate Change (2013b) on the expected price effect of the combined RO and Contract for Difference Feed-in Tariff. These are expressed in 2012 terms on a per-MWh basis, which we uprate to 2013 values.

Table B.4 Real cost per unit of electricity of RO and CfD FITs in 2020, by end user

User	£/MWh, 2013 prices	Carbon tax (£/tCO_2e)
Households	14.45	**49.35**
Non-energy-intensive business	13.42	**45.83**
Energy intensive business	13.42	**45.83**

Source for costs: Department of Energy and Climate Change (2013b) uprated to 2013 values

[77] http://www.ofgem.gov.uk/Sustainability/Environment/RenewablObl/Documents1/Buy-out%20price%20and%20mututalisation%20ceiling%202013_14.pdf.

Carbon Reduction Commitment (CRC) Energy Efficiency Scheme

The CRC fixes a price of £12/tCO_2 for imputed emissions from gas and electricity use by some firms. The policy specifies conversion factors from gas and electricity use in MWh to assumed carbon emissions which then form the basis of the tax. We use these policy-specific conversion factors to convert the CRC tax rate to a per-MWh basis, and then use the conversion factors from Table B.1 to re-express the CRC as a carbon tax. We do this rather than just take the £12/tCO_2 figure directly because the CRC conversion factor for electricity is based on grid average emissions whereas we are interested in the marginal tax. In other words, a firm participating in the CRC that increased electricity consumption by 1MWh would pay a tax for that extra unit based on the policy-specific CRC conversion factor estimated on the basis of the average carbon content of electricity. However the carbon content of the marginal (CCGT) unit will be lower, and so the implicit marginal carbon tax higher, than the CRC rate.

The calculations and resulting carbon tax rates are shown in Table B.5.

Table B.5 Rates of CRC and implicit carbon prices, 2013–4

Fuel	**Tax rate**	**Policy-specific conversion factor**	**Energy-based tax rate**	**Implied marginal carbon tax**
Electricity	£12/tCO_2	0.541tCO_2/MWh	£6.49/MWh	**£17.64/tCO_2e**
Gas	£12/tCO_2	0.184tCO_2/MWh	£2.20/MWh	**£11.92/tCO_2e**

Source for policy-specific conversion factors: Environment Agency (http://publications.environment-agency.gov.uk/PDF/GEHO0312BWGE-E-E.pdf, Annex C).

By 2020 the CRC rate will have risen to £16/tCO_2, increasing the implied marginal carbon tax on gas proportionally. However, the effective carbon tax imposed on electricity will have fallen. We assume that the policy-specific conversion factor, currently 0.541 tCO_2/MWh, will fall in line with grid-average emissions for the covered sectors. We take the forecast grid-average emissions for the commercial and public sector in 2020, 0.196 tCO_2e/MWh (DECC and HMT, 2012, Table 1), as the future policy-specific conversion factor. We use this to calculate the energy-based tax rate, and then convert this into an implied marginal carbon tax using the 2020 carbon content of marginal electricity consumption from the same source.

Since in 2020 the grid-average emissions of carbon are around two-thirds the marginal rate, rather than 1.4 times the marginal rate as currently, this means the marginal carbon tax is likely to be *lower* than the headline rate.

See Box 6.1 and Appendix B.4 for more on the issue of marginal versus grid-average emissions.

Table B.6 Rates of CRC and implicit carbon prices, 2020

Fuel	Tax rate	Policy-specific conversion factor	Energy-based tax rate	Implied marginal carbon tax
Electricity	£16/tCO_2	0.196tCO_2e/MWh	£3.14/MWh	**£10.73/tCO_2e**
Gas	£16/tCO_2	0.184tCO_2e/MWh	£2.94/MWh	**£15.89/tCO_2e**

Source for policy-specific conversion factors: Environment Agency (http://publications.environment-agency.gov.uk/PDF/GEHO0312BWGE-E-E.pdf, Annex C).

Small-scale Feed-in Tariffs (FITs)

Households and firms installing various small-scale renewable energy technologies are eligible to receive subsidies known as Feed-in Tariffs. The costs to energy companies of paying these subsidies are recouped through increases in electricity bills for households and firms. This may be through increases in marginal electricity prices or increases in fixed costs (e.g. standing charges). We follow the assumption of the Department of Energy and Climate Change (2013b) analysis of the impact of policies on energy prices and bills that these costs are recouped through higher marginal prices and so treat them as imposing effective carbon taxes on electricity use. Department of Energy and Climate Change (2013b) estimate that, in 2012 prices, FITs will add £2/MWh to the electricity price for households and firms in 2013, and £5/MWh by 2020.[78] This includes a VAT rate of 5%. Uprating to 2013 prices using the RPI inflation rate as reported by Office for Budget Responsibility (2013), removing the VAT contribution, and using the conversion factor in Table B.1, this equates to a marginal carbon price of **£5.34/tCO_2e** for electricity in 2013, and **£16.79/tCO_2e** in 2020.

Warm Home Discount (WHD)

The WHD is a policy delivered through energy companies which offers certain households a discount of £135 on their electricity bill which is recouped through increases in the energy bills (electricity and gas) of all domestic customers. Again, we assume this results in higher prices at the margin and take Department of Energy and Climate Change (2013b) estimates of the effect as the basis for our figures, shown in Table B.7. The real price effect is assumed to be the same in 2020; however, the impact in carbon terms for electricity increases because of the reduction in marginal carbon content over time. Again, any effect of VAT is removed before calculating the carbon price.

Note that we do not treat people receiving the WHD rebate as receiving an effective subsidy in carbon terms, since in principle a bill rebate should not affect marginal incentives to use electricity.

[78] For large, energy-intensive firms, a range of £0 to £2/MWh in 2013 and £0 to £5/MWh in 2020 is given in the DECC estimates; for consistency across firms we assume £2/MWh and £5/MWh respectively.

Table B.7 Effect of WHD on domestic energy prices and effective carbon taxes, households

Year	Fuel	Impact on price (£2013/MWh)	Implied carbon tax (£/tCO_2e)
2013	Electricity	1.97	**5.34**
	Gas	0.39	**2.13**
2020	Electricity	1.97	**6.71**
	Gas	0.39	**2.13**

Source: Price effects from Department of Energy and Climate Change (2013b), uprated to 2013 values.

Energy Company Obligation (ECO)

ECO offers energy-efficiency improvements through various mechanisms to the domestic sector, delivered through energy companies. As with the WHD, the costs are recouped through domestic energy bills, and we follow Department of Energy and Climate Change (2013b) estimates of the impact on prices to calculate effective carbon prices. The results are shown in Table B.8.

Table B.8 Effect of ECO on domestic energy prices and effective carbon taxes, households

Year	Fuel	Impact on price (£2013/MWh)	Implied carbon tax (£/tCO_2e)
2013	Electricity	5.90	**16.02**
	Gas	1.97	**10.64**
2020	Electricity	6.88	**23.50**
	Gas	1.97	**10.64**

Source: price effects from Department of Energy and Climate Change (2013b), uprated to 2013 values.

Domestic energy VAT subsidy

Households face a reduced rate of VAT of 5% on energy use. This is an effective subsidy of 14.3% (1.20 ÷ 1.05) compared with the standard 20% VAT rate. Firms pay the standard rate on their energy use. To convert this to a carbon value, we need to know the price of domestic energy in order to establish the per-unit value of the subsidy and thus the implicit carbon price.

For electricity and gas in 2013 we take DECC estimates of the national average marginal price per kWh in 2012.[79] These are then uprated to 2013 prices using

[79] The size of the implicit subsidy will vary across households according to the actual marginal prices they pay on their individual tariffs. Electricity figures are available from https://www.gov.uk/government/uploads/system/uploads/attachment_data/file/172857/qep224.xls (we use the standard electricity figure). Gas figures are available from https://www.gov.uk/government/uploads/system/uploads/attachment_data/file/172861/qep234.xls.

the out-turn RPI inflation rates for electricity (7.6%) and gas (8.5%) between April 2012 and April 2013.[80]

For 2020 we take DECC estimates of the average electricity and gas prices per MWh (given in 2012 prices), provided in Table 3 of Department of Energy and Climate Change (2013b). These are uprated to 2013 prices using the headline RPI inflation rate as reported by Office for Budget Responsibility (2013).

For coal, we take ONS estimates of the retail price of a 50 kg bag of domestic coal in April 2013.[81] This was £16.73, or 33.5p/kg.

For LPG there are no available official statistics and so we use an internet search for the price of a 47 kg refill (commonly used for domestic heating).[82] This was £67.49, or £1.44/kg.

Table B.9 reports the figures and calculations for the size of the implicit subsidy on a carbon basis for each fuel.

Table B.9 Rates of implicit subsidy from reduced-rate VAT on domestic energy

Fuel	Price (2013 base year)	Effective subsidy (14.3% of price)	Unit	Carbon tax (£/tCO_2e)
Electricity 2013	135.84	19.41	£/MWh	**-52.73**
Gas 2013	40.99	5.86	£/MWh	**-31.68**
Electricity 2020	213.05	30.44	£/MWh	**-103.96**
Gas 2020	56.42	8.06	£/MWh	**-43.61**
Coal 2013	334.60	47.80	£/tonne	**-16.97**
LPG 2013	1435.60	205.14	£/tonne	**-69.95**

Sources as specified in the text.

[80] Data from Office for National Statistics (http://www.ons.gov.uk/ons/datasets-and-tables/data-selector.html?table-id=2.2&dataset=mm23), series CZCZ and CZDA.

[81] See http://www.ons.gov.uk/ons/datasets-and-tables/data-selector.html?table-id=3.1&dataset=mm23, series CZMO.

[82] http://www.fuels4u.com/default/gas/calor-gas-and-appliances.html, accessed 20 June 2013.

B.2 Policies affecting different end users and fuels

Table B.10 summarises which policies affect which end-user/fuel combinations.

Table B.10: Policy coverage for all end-user and fuel type combinations

	Households	Small business	Medium business	Large energy-intensive business
Gas	VAT subsidy ECO WHD	CCL	CCL CRC	CCA rate of CCL
Electricity	ROC ETS CPSR VAT subsidy ECO WHD FITs	ROC ETS CPSR CCL FITs	ROC ETS CPSR CCL CRC FITs	ROC ETS CPSR CCA rate of CCL FITs
LPG	VAT subsidy	CCL	CCL	ETS CCA rate of CCL
Coal	VAT subsidy	CCL	CCL	ETS CCA rate of CCL

B.3 Carbon prices for LPG and coal by end user and policy

Figures B.1 and B.2 show carbon prices by end user for LPG and coal respectively. They follow Figures 6.2 and 6.3, which looked at electricity and gas, breaking down the carbon price for each policy individually and giving the overall price as a black dot.

Note that fewer policies apply directly to LPG and coal than is the case even for gas. Firms do not pay the CRC for these fuels, and none of the price increases levied by suppliers to recoup the costs of FITs, WHD or ECO apply to non-metered fuels.

For households it is *only* the implicit subsidy from the reduced rate of VAT which affects carbon prices for LPG and coal. The LPG subsidy is larger in carbon terms than the coal subsidy. LPG is more expensive per kilo, and since the VAT reduction is a subsidy proportional to the price this leads to a larger implicit carbon subsidy compared with coal (which is not fully offset by the higher carbon content of LPG relative to domestic coal).

For firms it is only the CCL (or CCA rate of CCL) and, for those large enough to be direct participants, the ETS which impose implicit carbon prices on LPG and coal. For large, energy-intensive firms, participation in the ETS roughly offsets the reduced rate of CCL they pay on these fuels, leading to similar implicit carbon prices across different firm types. For coal, for example, small and medium sized firms just pay the CCL, equivalent to £6.54/tCO_2e. Large energy-intensive firms pay the reduced CCA rate (£2.29/t) but also the ETS (£3.50/t) giving a combined carbon price of £5.59/t.

Figure B.1: Implicit carbon prices (£/tCO_2e) for LPG use, by end user (2013–4)

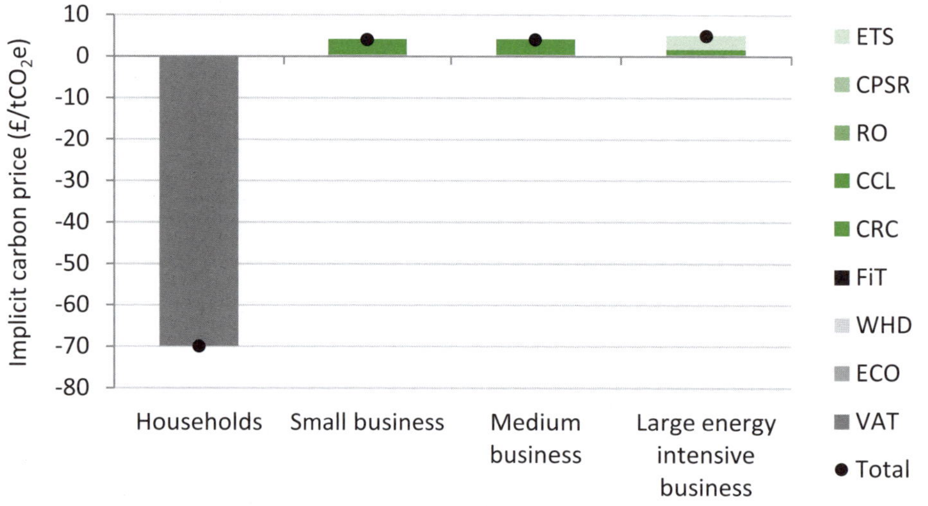

Source: Authors' calculations, based on assumptions and sources in this Appendix.

Figure B.2: Implicit carbon prices (£/tCO_2e) for coal use, by end user (2013–4)

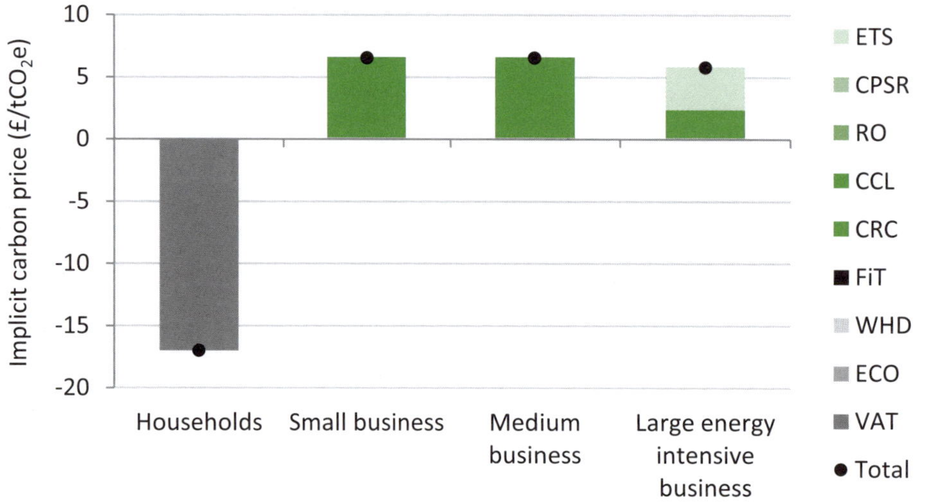

Source: Authors' calculations, based on assumptions and sources in this Appendix.

Figure B.3 shows the overall carbon prices across all four fuels for each end user, extending the analysis of Figure 5.3 to include these non-metered fuels as well.

Carbon prices for non-metered fuels are typically lower than for metered fuels. Note the implicit subsidy for LPG for households from the reduced VAT rate is much larger than for any other fuel.

Figure B.3: Implicit carbon prices (£/tCO_2e), by end-user and fuel type (2013–4)

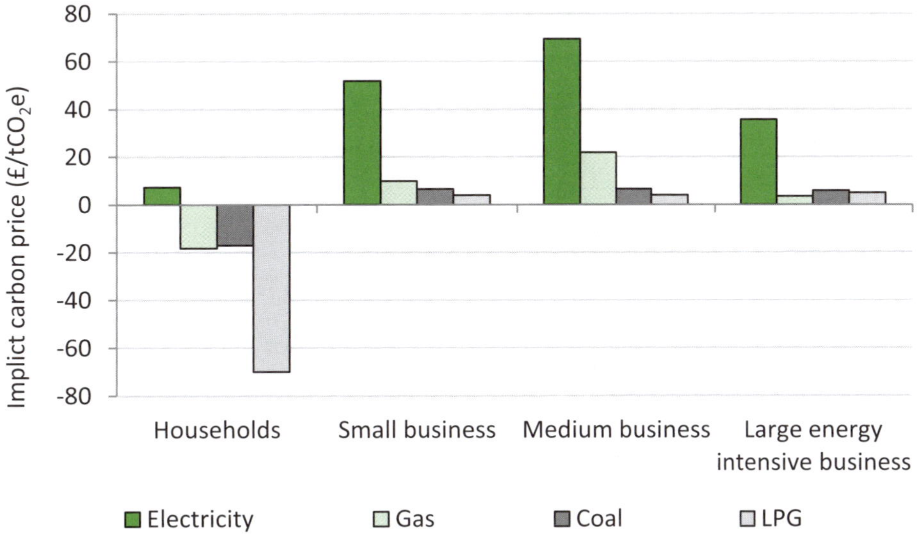

Source: Authors' calculations, based on assumptions and sources in this appendix.

B.4 Carbon prices using grid-average electricity emissions

As described in Chapter 6 and at the beginning of this appendix, an alternative question about variation in carbon prices for fuel use can be answered by using grid-average emissions from electricity use to convert taxes on electricity use to carbon equivalent terms rather than the emissions of the marginal plant.

Table B.11 shows the figure for grid average emissions. It is higher than the marginal factor used so far, since the mix of fuels also includes relatively polluting coal-fired generation. This will reduce implicit carbon prices on electricity: since a given equivalent tax per MWh of electricity use contains more carbon on average (494 gCO_2) than at the margin (368 gCO_2), using the average emissions factor will reduce implicit carbon taxes on electricity use.

Table B.11 Grid average emissions factor

Fuel	Units	Factor	Notes and source(s)
Electricity (average)	tCO_2e/MWh	0.494	Department for Energy and Climate Change and HM Treasury (2012) Table 1.

Note that we also adjust the methodology for calculating implicit prices for the CPSR and the RO. The CPSR only applies to electricity generated by gas and coal (i.e. non-nuclear and non-renewable electricity). In 2011 these accounted for 69.3% of generation (Department of Energy and Climate Change, 2012o) and so we multiply the CPSR (£4.94/tCO_2e) by 69.3% to get the effective carbon tax on electricity use (£3.42/tCO_2e). Similarly, the RO only applies to non-renewables (but including nuclear) which accounted for 90% of generation in 2011 (Department of Energy and Climate Change, 2012o). Thus we multiply the per-MWh price of the RO as calculated above by 90% and then convert it to carbon equivalent using the grid-average emissions. This gives an implicit carbon price for the RO here of £15.70/tCO_2e on electricity.

Figure B.4 shows the electricity and gas carbon prices for each end user using the grid average methodology for electricity (this compares directly to Figure 6.4), and Table B.12 shows the electricity carbon tax for each user on the two different definitions.

Figure B.4: Implicit carbon prices (£/tCO_2e), by end user (electricity based on average carbon content), 2013–4

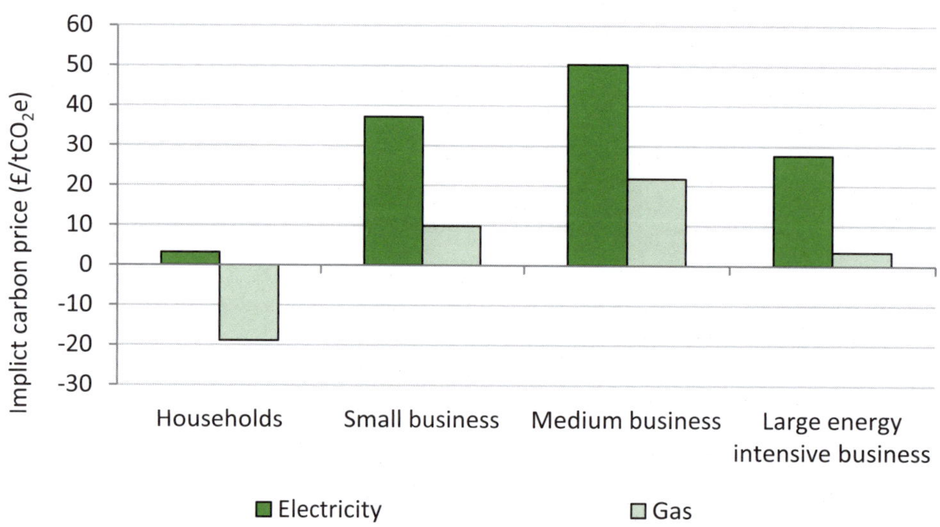

Source: Authors' calculations, based on assumptions and sources in this appendix.

As expected, the average-based electricity carbon taxes are lower than the marginal, though still considerably higher than those on gas for each end user, and still with variation in the tax rate across end users. Indeed, on an average basis the relative dispersion across users widens: medium-sized businesses (in the CRC) pay around 10 times the implicit carbon tax rate for electricity use than households under the marginal definition, but 15 times as much under the average definition. Thus, the broad conclusions remain unchanged whether the analysis is based on the marginal electricity unit or the average.

Table B.12: Implicit carbon prices (£/tCO_2e) on electricity use by end user, marginal and average definition of electricity carbon content

End user	Marginal	Average
Households	5.92	3.22
Small business	51.53	37.22
Medium business	69.17	50.37
Large energy-intensive business	38.71	23.68

Source: Authors' calculations, based on assumptions and sources in this appendix.